Side Effects of Chemotherapy and Radiotherapy

Jiajiu Shaw, Ph.D.
Frederick Valeriote, Ph.D.
Ben Chen, Ph.D.

DISCLAIMER

The contents provided in this book are presented to educate readers about the side effects of chemotherapy and radiotherapy. This book does not constitute the practice of any medical, nursing or other medical professional health care advice, diagnosis, or treatment. It is written solely for informational and educational purposes; nothing in the book should be used as a substitute for medical advice, diagnosis, or treatment.

Although some commercial products are mentioned in this book, it does not mean that we endorse the specific products. None of the contents in this book represents or warrants that any particular product or service is appropriate or effective for the readers. Readers shall always seek the advice of their physicians or other medical practitioners with any questions regarding personal health or medical conditions.

Side Effects of Chemotherapy and Radiotherapy
Published by Shaw Creative Inc.
Henderson, Nevada

Illustrations by Cindy Shaw

ISBN 978-0-9904226-4-8

Printed in the United States of America

Preface

Cancer is a group of diseases with similar characteristics, out-of-control growth of cancer cells. Conventional cancer treatments include surgery, chemotherapy, and radiotherapy. Although technology advancements have made vast improvements in the last two decades or so, there is no perfect treatment for cancer because of its complexity; one of the major issues related to cancer treatment is the unwanted side effects.

In this book, we will focus on the side effects induced by chemotherapy and radiotherapy. We will briefly introduce chemotherapy and radiotherapy; we will also describe the side effects, current methods to tackle the side effect issue, and how to manage the side effects.

The purpose of this book is to help people understand the potential side effects of chemotherapy and radiotherapy. In addition, we hope that this book can help readers understand how to recover faster from chemotherapy and/or radiotherapy and reinvigorate their health. However, it is important for the readers to know that this book is not intended to substitute the expertise and judgment of your physicians or other medical practitioners.

Because of the technical aspects of this book, it is recommended to be used as a Cancer 101 text book. Nevertheless, people with some technical background in chemistry/biochemistry should be able to read through this book without any difficulty.

Finally, if this book is useful for you in any way, please share it with your loved ones and friends.

Jiajiu Shaw, Ph.D.
Frederick Valeriote, Ph.D.
Ben Chen, Ph.D.

Table of Contents

Part I

Chemotherapy and Its Side Effects

Part I

Chapter 1
Introduction

Mutation is not uncommon in our daily lives. Most mutated cells either keep their normal cellular functions or undergo a programmed cell death (apoptosis) and die at some point in time. In addition, many mutated cells are recognized and rapidly removed by the body's immune system. Only a very small portion of mutated cells will evolve further and become cancerous cells with uncontrollable proliferation; this behavior is different from that of normal cells. For adult, most normal cells are well differentiated except for certain types of cells, such as epithelial cells and bone marrow hematopoietic progenitor cells, which are capable of continuous proliferation; unlike cancerous cells, the growth of normal cells is highly regulated and controlled.

Cancer is a group of various diseases in that some cells are mutated in certain ways so that they grow and divide continuously. Currently, there are more than 200 known cancers that afflict humans.

For the treatment of cancer, there are generally three conventional strategies: surgery, chemotherapy, and radiotherapy. In this book, we will focus on chemotherapy, radiotherapy, and their side effects.

Chemotherapy is a systemic treatment of cancer patients with one or more anticancer drugs (chemical agents) to kill cancer cells and/or slow down their growth. The name "chemotherapy" comes from **chem**ical agents being used as a **therapy** for treating cancer. Chemotherapy plays a critical role in cancer treatment; it is estimated that more than half of all cancer patients receive chemotherapy.

Depending on the dosage form, cancer drugs may be administered through several different routes, including intravenous injection/infusion, oral (mouth) delivery, or a slow release dosage form embedded in a tissue by implantation. A chemotherapy regimen (a treatment plan including the dosing and schedule) usually includes the administration of at least one drug to fight cancer at a predetermined dose according to a suitable schedule, usually daily, weekly, or every several weeks. Due to the different characteristics of individual cancer drugs, the chemotherapy regimen for each drug may be different. Because cancer treatment is a delicate process, it is important to know that staying on the chemotherapy regimen gives a patient the optimal chance for a successful result.

Many times, additional supplementary drugs are used simultaneously in a chemotherapy regimen in order to achieve a better outcome to either increase the anticancer potency or reduce the side effects of chemotherapy; these drugs are often called adjuvant medications. Adjuvant drugs that are used to reduce the side effects of cancer therapy, including chemoprotective and radioprotective agents, will be discussed in Part III.

Most cancer drugs work by killing cancer cells or slowing down the growth of cancer cells. Unfortunately, most of the cancer drugs are not able to clearly distinguish between cancer cells and normal cells. Therefore, most cancer drugs will damage not only the fast-growing cancer cells but also the fast-growing normal cells, which include bone marrow cells, crypt epithelial cells (lining of small intestine), and hair follicles. In addition, a number of organs (e.g., liver, kidney, and heart) are often injured by chemotherapy.

In order to have a better understanding on how anticancer drugs work and what kind of side effects they can cause, let's take a look at some anticancer drugs in different categories.

Part I

Chapter 2
Anticancer Drugs

Chemotherapy added a major new capability to the treatment of cancer beginning with the use of alkylating agents and anti-folates in the 1940s; it continues to expand today with anticancer agents specifically targeting single subcellular structures such as proteasomes or large enzyme groups like the kinase inhibitors and anti-metabolite agents.

In addition, medical oncologists who deliver these chemical agents have become an important leg of the present therapeutic triad, which also includes surgeons and radiotherapists. This triad defines the protocol for today's impressive treatments.

For now, let's focus on chemotherapeutic drugs. In addition to the conventional anticancer drugs, in the last twenty years or so, many novel anticancer drugs have been developed and approved by the Food and Drug Administration (FDA).

While chemotherapy drugs may be categorized in different ways. In this book, we attempt to divide anticancer drugs into several categories according to Table 1.1 although some drugs could be placed under several categories. Another type of therapy, gene therapy, will not be discussed here.

Based on the listing in Table 1.1, a number of representative anticancer drugs will be described and discussed in the following sections.

Table 1.1. Categories of chemotherapeutic drugs

Category	Mechanism	Example
A. Alkylating agents	Affect DNA or cell division	nitrogen mustard, cyclophosphamide, nitrosourea, cisplatin, oxaliplatin
B. Antimetabolites	Affect DNA or cell division	methotrexate, 5-fluorouracil
C. Mitotic inhibitors (Tubule inhibitors)	Affect cell division	paclitaxel, vinblastine
D. Antibiotics (Anthrocyclines)	Affect DNA or cell division	doxorubicin, dactinomycin, bleomycin
E. Kinase modulators	Modulate cell behaviors	imatinib
F. Hormones	Modulate cell behaviors	tamoxifen
G. Monoclonal antibodies	Modulate cell behaviors	cetuximab, bevacizumab
H. Other approaches	Modulate cell behaviors	antibody + chemotherapy

A. Alkylating agents

A classical alkylating agent works by attaching an alkyl group or a small molecule to hinder/stop the replication of DNA in cancer cells. Classic alkylating agents include nitrogen mustard, and cyclophosphamide. Platinum drugs are a different type of alkylating agent. Instead of directly attaching an alkyl group onto DNA, a platinum drug forms coordination bonds with DNA and hampers the replication of DNA. Some examples of alkylating agents are described below.

A.1. Nitrogen mustards

One of the first alkylating agents was nitrogen mustard, NH2 (Fig. 1.1). Nitrogen mustards are alkylating agents but are not specific in their binding site. The utility of nitrogen mustards is limited because they are very reactive and consequently very short lived.

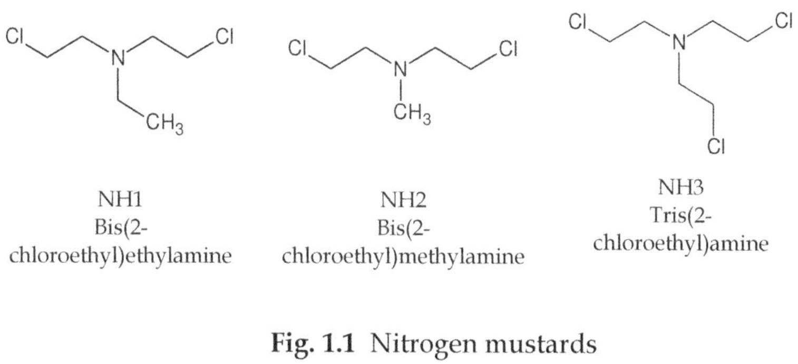

NH1 Bis(2-chloroethyl)ethylamine	NH2 Bis(2-chloroethyl)methylamine	NH3 Tris(2-chloroethyl)amine

Fig. 1.1 Nitrogen mustards

Because of its excessive toxicity, the original nitrogen mustard, NH2, is not often used nowadays. Instead, some of its analogs or derivatives, such as cyclophosphamide, are being used.

A.2. Cyclophosphamide

Cyclophosphamide (Fig. 1.2) is a very old anticancer drug developed in the 1950s. It is an alkylating agent that may be categorized as a nitrogen mustard.

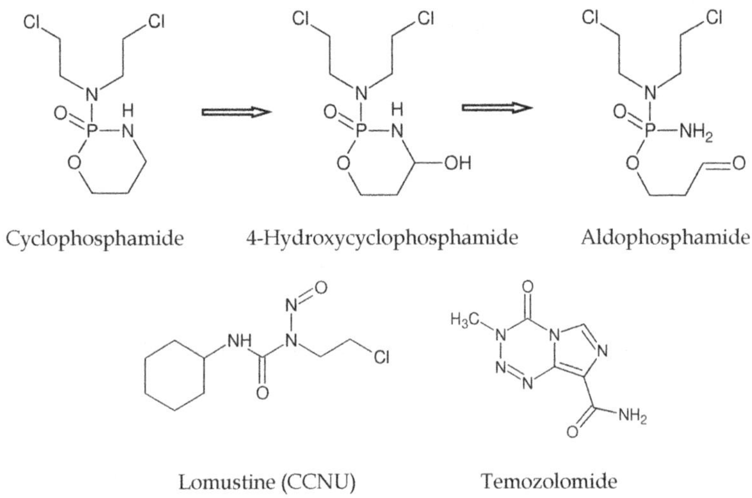

Fig. 1.2 Cyclophosphamide and related compounds

Although cyclophosphamide (Fig. 1.2) is an analog of nitrogen mustard, it is a prodrug and inactive as opposed to nitrogen mustard. Cyclophosphamide is a prodrug (i.e., it is metabolized and become active in the body); it is metabolized in the liver by oxidase enzymes to become 4-hydroxycyclophosphamide, which co-exists with its tautomer, aldophosphamide (Fig. 1.2).

Cyclophosphamide is administered by either *i.v.* injection or oral administration (25 and 50 mg tablets). It is often used in combination with other anticancer drugs to treat solid tumors in the lymphatic cells of the immune system (lymphomas) and some other types of cancers. In addition to its anticancer effect, cyclophosphamide also suppresses the immune system and is also referred to as immunosuppressive.

Potential side effects of cyclophosphamide include bone marrow suppression (causing leucopenia, anemia, and thrombocytopenia), nausea, vomiting, diarrhea, jaundice, and alopecia. Cyclophosphamide may cause kidney failure, and it also may affect the heart and lungs. Cyclophosphamide may also cause inflammation of the urinary bladder with bleeding (hemorrhagic cystitis). This can result in lower abdominal pain, problems urinating due to blood clots, and anemia due to loss of blood.

A.3. Nitrosoureas

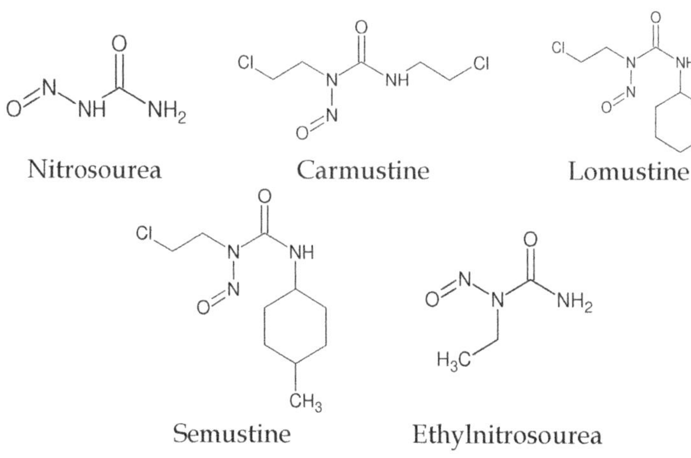

Fig. 1.3 Examples of nitrosoureas

Nitrosoureas (Fig. 1.3) are another group of alkylating agents. Because of their lipid solubility, they form the basis for brain tumor treatment as they are able to cross the blood-brain barrier. Nitrosoureas can be used alone or used in a newer technological product, Gliadel wafer [1].

A.4. Cisplatin and other platinum drugs

Cisplatin, *cis*-dichlorodiammine platinum(II) (Fig. 1.4), is the grandparent of all platinum drugs. It is a widely prescribed anticancer drug with therapeutic activity against various solid tumors, but also with substantial side effects. The anticancer activity of cisplatin was discovered by Dr. Barnett Rosenberg of Michigan State University in 1965 [2]. In 1978, the first formulated pharmaceutical product of cisplatin, Platinol, was approved by the FDA and sold by Byistol-Myer Squibb.

As an alkylating agent, the platinum ion of cisplatin forms coordination bonds mostly with guanine on DNA. This crosslinking causes the interference of DNA replication, thus, inhibits the replication of cancer cells.

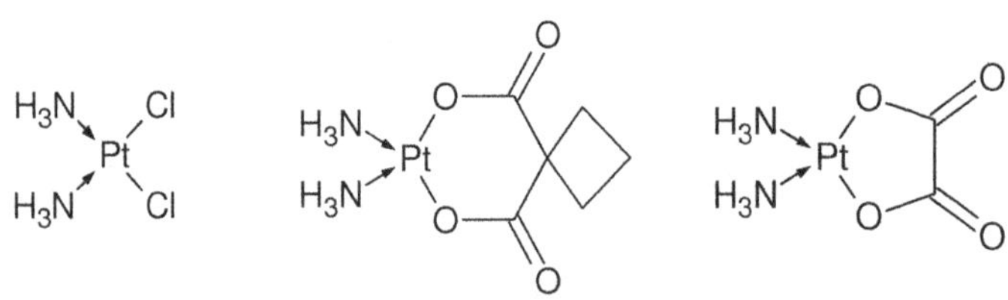

Cisplatin Carboplatin Oxaliplatin

Fig. 1.4 Cisplatin, carboplatin, and oxaliplatin

Since the discovery of cisplatin's anticancer activity, thousands of cisplatin analogues have been made and/or studied. Two of the notable platinum drugs successfully developed by Bristol-Myers Squibb are carboplatin and oxaliplatin (Fig. 1.4); both were approved by the FDA for treating a variety of cancers.

In the last thirty years or so, the impact of platinum drugs on cancer treatment has been tremendous and has benefited many cancer patients. However, because platinum drugs are often associated with significant side effects, scientists are still working on finding better platinum compounds with fewer/less side effects. Currently, there are a number of new platinum drugs under investigation or development, including some of the platinum compounds synthesized by one of the co-authors, Dr. Jiajiu Shaw [3, 4]. A couple of these new platinum compounds are shown in Fig. 1.5. The dinuclear platinum compound is the first to show a coordination bond between the amid oxygen and platinum(II) [5].

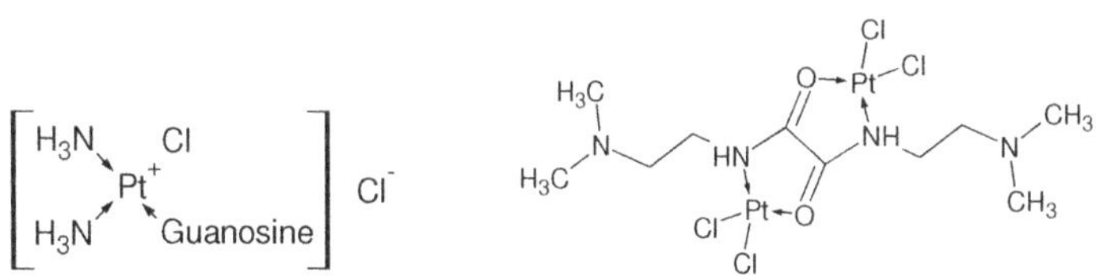

Fig. 1.5 Examples of a 3-Nitrogen Pt compound (left) and a dinuclear Pt compound (right)

B. Antimetabolites

An antimetabolite is a drug that interferes with normal cell metabolism by inhibiting certain metabolite pathways and stopping cell growth or cell division [6]. The molecular structure of an antimetabolite is usually similar to the metabolite that they interfere with. They usually induce cell death during the S phase of the cell cycle.

For example, folic acid is an important component in the synthesis and repair of DNA and RNA. An analog of folic acid, methotrexate(Fig. 1.6), has a structure that is almost identical to that of folic acid but acts very differently; methotrexate inhibits cell division to achieve its anticancer effect.

B.1. Methotrexate (also referred to as amethopterin)

Methotrexate has been used for treating cancer and autoimmune diseases for many years. Although it has been off-patent for many years, it is still very highly prescribed.

Methotrexate is an antimetabolite that interferes with the use of folic acid. As shown in Figure 1.6, the structures of methotrexate and folic acid are strikingly similar. Because folic acid is essential for the synthesis and repair of DNA, especially during cell growth and division, it is logical that methotrexate interferes with the growth of cancer cells.

Methotrexate is cytotoxic and has a wide range of potential side effects including myelosuppression and mucositis. Methotrexate is also highly teratogenic and should not be taken by women during pregnancy or expected to be pregnant. It is generally true that no chemotherapy is given for pregnant women and, for women of child-bearing age, precaution must be taken.

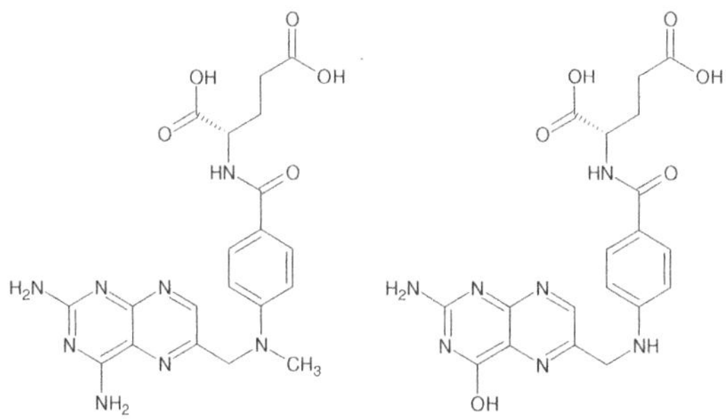

Fig. 1.6 Structures of methotrexate (left) and folic acid (right)

Sometimes, folic acid is used to reduce the side effects of methotrexate. However, folic acid could act as an antidote to methotrexate, therefore, it is better to use folic acid on the off days during methotrexate treatment [7].

B.2. Fluorouracil (5-Fluorouracil or 5-FU)

Fluorouracil (Fig. 1.7) is sold under different trade names such as Adrucil and Efudex. It is classified as an antimetabolite and has been used for treating cancer for many years. Due to its similarity to thymine, 5-FU interrupts the synthesis of thymidine, one of the building blocks of DNA. As a result, DNA replication is interrupted and the growth of cancer cells is hampered.

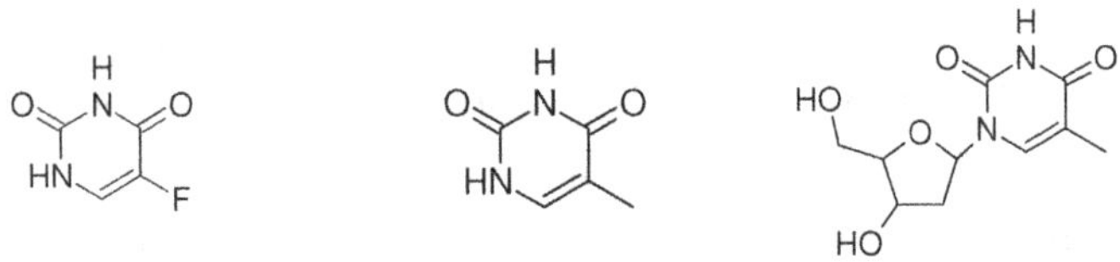

Fig. 1.7 5-FU (left), thymine, and thymidine (right)

Fluorouracil has been used mainly in treating colon cancer and has been used in combination with other drugs, including leucovorin, carboplatin, and paclitaxel. It has also been used in other solid tumors including pancreatic cancer and breast cancer. Side effects of fluorouracil include myelosuppression, dermatitis, and mucositis.

C. Mitotic inhibitors (Tubule inhibitors)

Mitosis is one step of the cell cycle (Fig. 1.8) in which a cell is divided into two identical cells and each has its own nucleus. A drug that inhibits mitosis (cell division) is referred to as a mitotic inhibitor, also referred to as a tubulin inhibitor. Because microtubules play a very important role in mitosis, disruption of microtubules disrupts cell division thus achieving the anticancer effect.

Tubule inhibitors, including alkaloids and terpenoids, are usually derived from plants. They usually work by interfering with microtubule function or inhibiting the assembly of microtubules; microtubules are special proteins in tubular shape that are important components of cytoskeleton, which is a dynamic skeleton in cytoplasm that maintains cell shape and plays an important role in cellular division. Examples of this type of drug include paclitaxel and vinblastine.

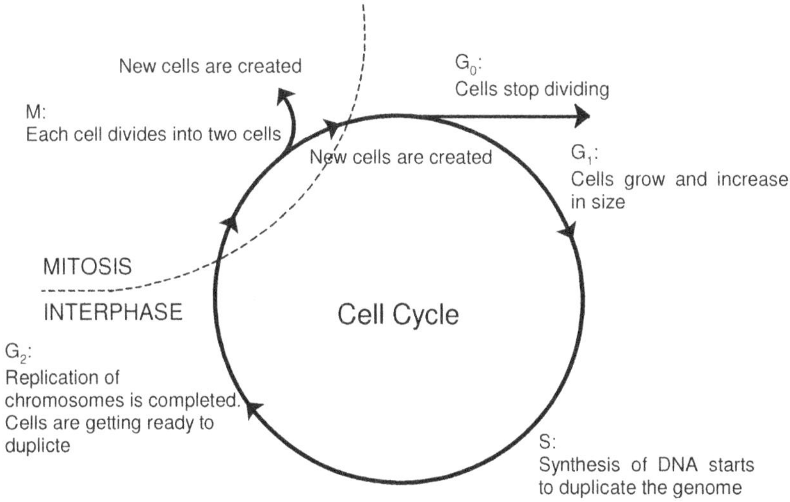

Fig. 1.8 Cell cycle

C.1. Paclitaxel (Sold as Taxol®)

Paclitaxel (Fig. 1.9) was isolated from the bark of the Pacific yew, Taxus brevifolia, by Monroe Wall and Mansukh Wani in 1967. It was co-developed by NCI and Bristol-Myers Squibb. The drug product was approved as Taxol® by the FDA in 1992. Since its approval, Taxol has been used to treat patients with ovarian, lung, and breast cancer for a number of years.

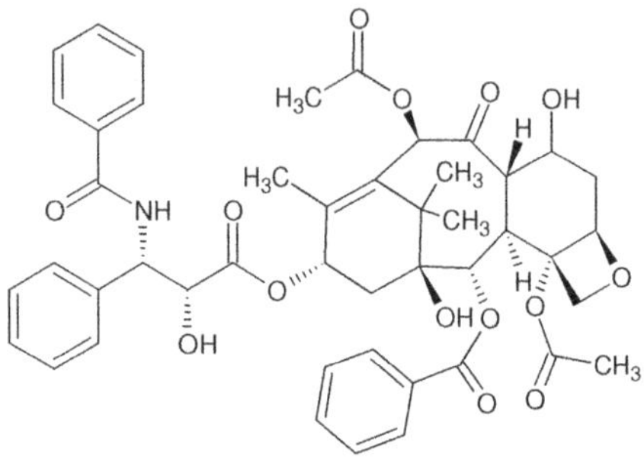

Fig. 1.9 Structure of paclitaxel

Paclitaxel works by hyper-stabilizing microtubules, major components of the dynamic cellular skeleton (cytoskeleton), to interfere with the normal breakdown of microtubules in the M phase of cell cycle (mitosis). In eukaryotic cells (cells with a nucleus), the cell cycle can be divided in two brief periods: interphase and the mitosis. During the interphase (I), the cell grows, accumulating nutrients and duplicates its DNA. During the mitosis phase (M), the cell splits itself into two distinct daughter cells. As a result of paclitaxel treatment, which hyper-stabilizes microtubules and hampers the mitosis, cancer cells are not able to replicate.

Similar to all other anticancer drugs, paclitaxel has a number of side effects. Serious side effects include unusual bleeding, skin rash, change in bowel habits, fever, chills, cough, difficulty swallowing, dizziness, shortness of breath, as well as severe exhaustion.

C.2. Vinblastine

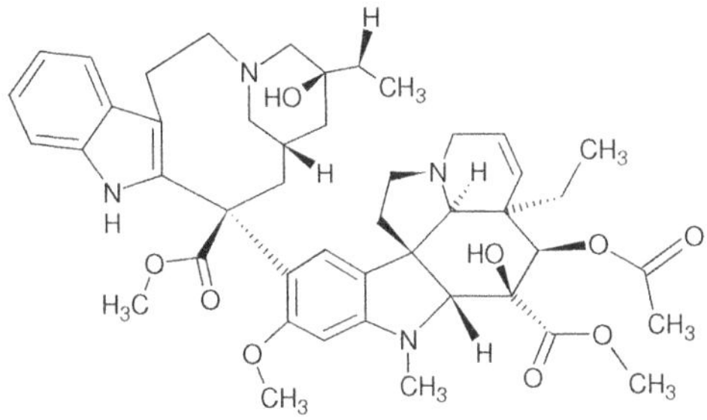

Fig. 1.10 Vinblastine

Vinblastine (Fig. 1.10) was first isolated from a Madagascar plant, periwinkle. Vinblastine has been used to treat Hodgkin's lymphoma, lung cancer, and bladder cancer [8-10]. It works as a mitotic inhibitor, similar to Paclitaxel through microtubule modulation by suppressing microtubule dynamics and the reduction microtubule mass.

D. Antibiotics

There are a number of antibiotics being used as anticancer drugs. Examples of antitumor antibiotics include doxorubicin, dactinomycin, epirubicin, and bleomycin. DNA binding agents can have a number of outcomes which result in cell death. They usually are planer in structure and multi-ringed to allow them to non-covalently bind to the bases in DNA (and RNA). Compounds like actinomycin D turn off RNA syntheses once they intercalate with the DNA.

Compounds, like bleomycin, create free radicals after intercalating with DNA strands.

In this section, we will focus on a class of anticancer drugs, anthracyclines. Anthracyclines are derived from Streptomyces bacteria. They are used to treat a number of cancers, including breast, ovarian, and lung cancer as well as leukemia. Anthracyclines are known to form intercalation with DNA to achieve anticancer effect although the complete mechanism is somewhat complex (intercalation is a reversible inclusion between two different molecules or groups). Anthracyclines have played a very important role in chemotherapy and have been used in many different types of cancers.

Although anthracyclines are effective against more cancers than most other classes of anticancer drug, anthracyclines do have several adverse effects including cardiotoxicity. A specific example of anthracyclines, Doxorubicin, is described below.

<u>Doxorubicin (trade names: Adriamycin PFS, Adriamycin RDF, or Rubex)</u>

Doxorubicin (Figure 1.11) is closely related to another anthracycline occurred naturally, daunorubicin. Just like all anthracyclines, doxorubicin intercalates with DNA and causes structural distortion of DNA. As a result, replication of DNA is inhibited and cancer cells eventually die. However, this intercalation with DNA can also happen to normal cells and causes significant side effects.

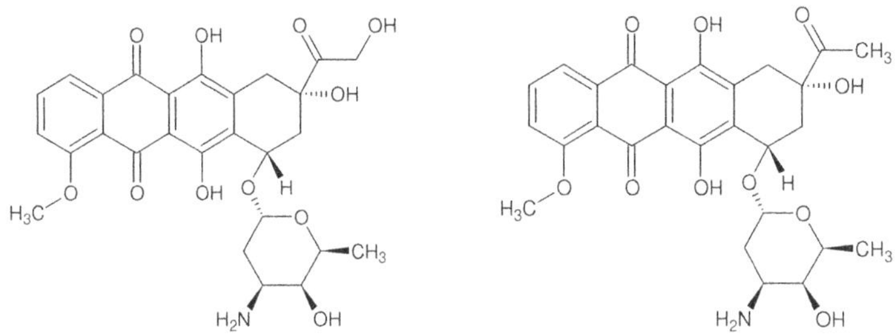

Fig. 1.11 Doxorubicin (left) and daunorubicin (right)

Doxorubicin is commonly used in the treatment of a wide range of cancers, including breast, ovarian, and lung cancer as well as leukemia. Doxorubicin is administered intravenously and its main side effects include cardiomyopathy, myelosuppression, and secondary malignancies.

E. Kinase Modulators

A protein kinase modifies a specific protein (substrate) by chemically adding a phosphate group (from ATP) to change the function/activity of the substrate that regulates the cellular pathways, especially signal transduction pathways. This process of adding a phosphate group to a substrate is referred to as phosphorylation. During the phosphorylation process, ATP loses a phosphate group and becomes ADP.

Phosphorylation is a very important reaction, which regulates a lot of biological processes. The phosphate groups are often added to tyrosine, threonine, or serine. As a result of the phosphorylation, many biological processes are regulated. Many diseases are due to the hyperactivity of protein kinases. Therefore, it makes sense to tackle these diseases by modulating the protein kinases.

More than 500 protein kinases have been identified in humans. For example, tyrosine kinase is an enzyme that helps attach a phosphate group onto a tyrosine residue in a protein; it is important in signal transduction and is implicated in the treatment of cancer.

Kinase modulators target a molecular abnormality by modulating the activity of certain kinase. Kinase modulators have been used in certain types of cancer. Examples of kinase modulators are shown below.

E.1. Imatinib mesylate (sold as Gleevec® by Novartis)

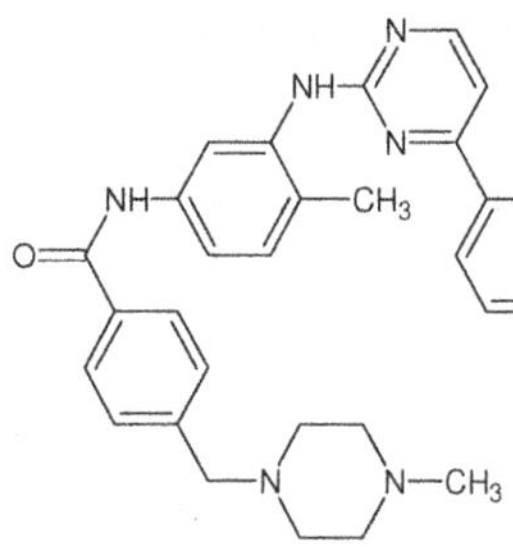

Fig. 1.12 Structure of imatinib

Imatinib (Figure 1.12) was first approved by the FDA in 2001 and sold as Gleevec®. It works by specifically targeting, and turning off constitutively active tyrosine kinases (Bcr-abl) which help cancer cells grow and multiply. Bcr-abl kinase is a hybrid product of a chimeric Bcr-abl oncogene caused by translocation between chromosomes 9 and 22. The tyrosine kinases cause several cancers including Philadelphia chromosome positive chronic myeloid leukemia (Ph+ CML) and gastrointestinal stromal tumor (GIST). Gleevac comes in tablet form. When given by oral administration, the active ingredient, imatinib, is rapidly absorbed.

Potential side effects of Gleevec include fluid retention, rash, nausea & vomiting, muscle cramps, and diarrhea. Detailed information can be found at www.gleevec.com.

E.2. Afatinib (Gilotrif™, Giotrif®)

The commercial product of afatinib (Fig. 1.13) is an orally administered inhibitor of the ErbB family of tyrosine kinases [11]. Afatinib also inhibits human epidermal growth factor receptor 2 (Her2) and epidermal growth factor receptor (EGFR). The inhibition is through the formation of a C-S covalent bond between afatinib and a cysteine in EGFR.

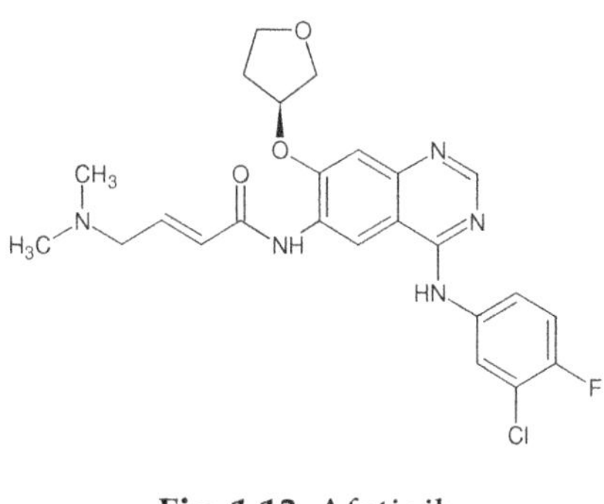

Fig. 1.13 Afatinib

Afatinib (Fig. 1.13) was first indicated for treating non-small cell lung cancer [12, 13]. In addition, it has been reported that afatinib may be useful to treat other cancers [14].

F. Hormones

Hormones are compounds internally secreted by a specific tissue/organ, transported by body fluids to another tissue/organ (receptor), and affect the function of the receptive tissue/organ. Most hormones exist naturally but some hormones can be synthesized.

For example, tamoxifen is a common hormone treatment for breast cancer. Also, a phase II clinical study suggested that the concomitant administration of melatonin (a pineal hormone) and tamoxifen may induce tumor regressions in metastatic breast cancer patients refractory to tamoxifen alone [15].

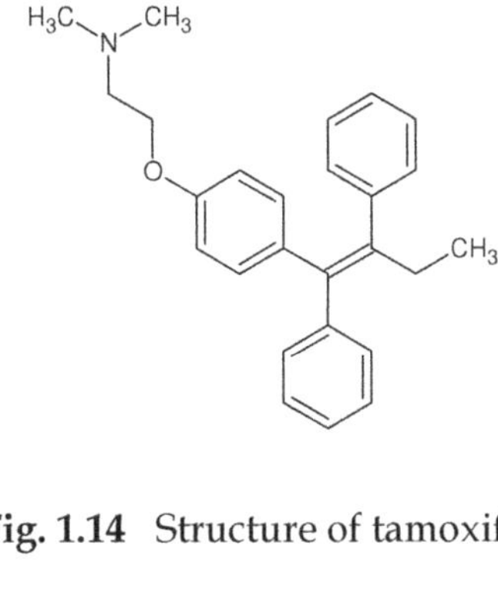

Fig. 1.14 Structure of tamoxifen

Certain breast cancer cells require the activation of estrogen receptor to grow. Tamoxifen (Fig. 1.14) is first metabolized by human cytochrome P450 and its metabolites then compete with estrogen in binding to and inactivating estrogen receptor, thus hampering the growth of the breast cancer cells [16].

Tamoxifen was approved by the FDA in December 1977. It is currently used for the treatment of both early and advanced estrogen receptor positive (ER+) breast cancer patients [17]. In June 1990, tamoxifen was approved by the FDA for an additional use to help prevent the recurrence of breast cancer wherein the cancer cells have not spread to the lymph nodes under the arm [18].

G. Monoclonal antibodies

Monoclonal antibody (mAb) plays a critical role in recognizing foreign antigen and then either stimulating or inhibiting immunological or cellular responses. There are several ways that mAbs can be used for therapy. For example, cetuximab is a monoclonal antibody that binds to the extracellular domain of EGFR thus achieving its anticancer effect by inactivating EGFR.

G.1. Cetuximab (Erbitux, ImClone)

Mutation by chemicals, virus, radiation etc. could cause the over-expression of EGFR. Resulting in uncontrolled cell division and is associated with a number of cancers such as lung cancer, colon cancer, and glioblastoma multiforme (the most common and aggressive brain tumor).

Cetuximab is a recombinant mouse/human (chimeric) monoclonal antibody that binds to the extracellular domain of the EGFR, thereby blocking the activation of EGFR to achieve its anticancer effect. It is used for the treatment of EGFR-expressing metastatic colon cancer after failure of irinotecan (Irinotecan, a semisynthetic analogue of the natural alkaloid camptothecin, is a drug used for the treatment of colon cancer) and oxaliplatin-based regimens. Cetuximab is used in treating colon cancer patients, either alone or in combination with chemotherapy.

Cetuximab is administered by infusion and may cause serious infusion reactions including rapid onset of airway obstruction, hypotension, shock, loss of consciousness, myocardial infarction, or cardiac arrest. Cetuximab may also cause electrolyte depletion for months after initiation of erbitux therapy.

G.2. Bevacizumab (Avastin)

Angiogenesis is a normal physiological process related to the growth of new blood vessels from pre-existing vessels. Angiogenesis can be stimulated as a result of tumor growth. Therefore, modulating the abnormally high angiogenesis has become a strategy for anticancer research after Judah Folkman proposed the theory in 1971.

Bevacizumab is the first approved angiogenesis inhibitor in the U.S. It is a humanized monoclonal antibody that works by blocking vascular endothelial growth factor A (VEGF-A).

Bevacizumab, in combination with intravenous 5-FU-based chemotherapy, was approved by the FDA for first- or second-line treatment of patients with metastatic colorectal cancer in 2004. It was subsequently approved for lung cancer (2006) and breast cancer (2008).

Major side effects of bevacizumab includes hypertension, surgery wound healing problem, higher risk of bleeding, thrombocytopenia, gastrointestinal perforation, and nasal septum perforation. Other common side effects include lower white blood cell (WBC) count, vomiting, constipation, diarrhea, and hair loss.

H. Other Types of Anticancer Drugs

Generally speaking, drugs in categories A to D affect DNA synthesis and/or cell division; they can also attack fast growing normal cells, including bone marrow cells, hair follicle cells, and crypt cells as mentioned before. The drugs in categories E to G including kinase modulators, hormone treatment, and mAbs can modulate tumor cell behavior without directly attacking DNA in those cells.

Other than those in these A-G categories, a number of newer approaches have been developed or under development. For example, making antibody-drug conjugates (ADCs) is one of the newer approaches. In an ADC, a potent anticancer drug is covalently linked to an antibody (the whole mAb or a fragment of an antibody) corresponding to specific antigen differentially overexpressed in cancer cells. With this design, the conjugate is able to discriminate between cancer cells (target of the antibody) and normal cells; once the ADC hooks up with the cancer cell, the potent anticancer drugs is released and attack the cancer cell. The end result is that anticancer effect may be increased and the side effects may be reduced.

You may have noticed that naturally occurring compounds are listed in the above examples for cancer treatment. In fact, nature provides many other powerful ingredients that may be useful for treating cancer and many other diseases. One important point for the reader to keep in mind: no natural products/ingredients can be legally claimed as effective drugs to treat or cure any diseases unless approved by the FDA.

Part I

Chapter 3
Side Effects of Chemotherapy

Chemotherapy can result in a wide range of side effects; many of them are due to its adverse activity on the fast-growing normal cells of the body such as bone marrow cells, hair follicle cells, and crypt epithelial cells on the small intestine lining. Other effects are due to its adverse activities on vital organs such as liver, kidney, and heart. Although the side effects induced by individual drugs can be different from drug to drug, they can be generalized and described below.

A. Bone Marrow

Because most anticancer drugs tend to damage fast-growing bone marrow cells, which produce new blood cells, blood-related side effects are not unexpected as a result of chemotherapy. Basically, there are three major blood-related side effects resulted from bone marrow damage by chemotherapy; they are briefly described below:

A.1. Neutropenia

Neutropenia is defined as an abnormal decrease in the number of neutrophils, an abundant type of granular WBC that is highly destructive of microorganisms. A low WBC count may leave a patient vulnerable to bacterial infection and the chemotherapy regimen has to be put on hold until the WBC count is back to an acceptable level for the chemotherapy to resume. This is often a problem that interferes with the chemotherapy regimen.

A.2. Anemia

Anemia is defined as a deficiency of hemoglobin (Hb), which is an iron-containing protein that carries oxygen from the lungs to the body's brains, muscles, tissues, and other organs; oxygen is critical to the health of tissues and organs. Symptoms of anemia include light headedness, tiredness, or dizziness when a patient stands up suddenly from a crouching position.

A.3. Thrombocytopenia

Thrombocytopenia refers to the condition of lower than normal platelet count. Platelets play a fundamental role in hemostasis (a complex process that causes the blood to clot and bleeding to stop) and are a natural source of growth factors including platelet-derived growth factor (PDGF), which plays a significant role in the repair and regeneration of connective tissues. A low platelet count may cause a person to experience bruising more easily or excessive bleeding. Patients with acute myeloid or lymphocytic leukemia often experience low platelet count. This is especially true when patients with acute leukemia are treated with chemotherapy

All of these three side effects are related to bone marrow damage and they can be diagnosed through the complete blood count (CBC) which represents the three types of cells in blood.

Currently, there are several drugs approved by the FDA to reduce individual side effects resulted from bone marrow damage. These drugs will be discussed later. In addition, some natural products have been claimed to have certain protective effects on bone marrow damage. However, one needs to look into the scientific evidence before making any conclusion or decision.

B. Hair

One of the most noticeable side effects of chemotherapy is hair loss. Remember that chemotherapy damages not only the fast-growing cancer cells but also the fast-growing normal cells? Because cells in the hair follicles are fast growing, they are more easily attacked by many cancer drugs, thus causing hair loss.

Hair loss caused by chemotherapy is often referred to as alopecia. Chemotherapy-induced alopecia is a significant problem in clinical oncology. Fortunately, the hair loss is usually not permanent and often the hair will begin to grow back several months after chemotherapy ends.

C. Intestine (also see mucositis)

Chemotherapy often causes diarrhea. Crypt epithelial cells on the lining of the small intestine reproduce almost as fast as the bone marrow cells. Therefore, crypt cells are often easily damaged by chemotherapy.

Epithelial cells derived from crypt cells actively secrete electrolytes and play an important role in the re-absorption of water secretion in the intestines. During a normal digestive process, suitable amount of water are secreted into the lumen of the small intestine as part of the digestion process; most of the water is reabsorbed afterwards. Damage of crypt cells by chemotherapy disturbs the normal water-absorption process in the intestine, thus causing diarrhea.

D. Other Side Effects

The above side effects are directly related to the rapid growth of cells. There are other significant side effects that are not directly related to the fast growing of normal cells. These side effects include liver damage, kidney damage, pain, rash, constipation, nausea, vomiting, weight loss, ototoxicity, electrolyte abnormality, cardiotoxicity, and depression of the immune system. Among them, liver and kidney damages are most critical and are often the sources of other side effects. Some of these side effects are described below.

D.1. Liver damage

Liver is a vital organ and has a wide range of functions. Liver produces bile to aid in digestion through the emulsification of lipids and plays a major role in metabolism. Other functions of liver include hormone production, decomposition of red blood cells (RBCs), plasma protein synthesis, and detoxification of small and complex molecules.

Theoretically, all drugs are associated with certain toxicity (but to different extents) and liver is the major organ that is supposed to "detoxify" the drugs mainly by breaking down the drugs into metabolite(s). The other way for liver to make a drug less toxic is to make it more water soluble, thus more easily excreted out of the body and reduce potential toxicity. However, a small percentage of drugs are able to evade the "protection mechanism" by the liver. In this case, the drugs pass through the body's major detoxifying guard (liver), damage liver, and move to other organs to exert other side effects.

A typical example of anticancer drugs with potential liver toxicity is cisplatin. Tyrosine kinase inhibitors are also associated with potential liver toxicity. Recently, the FDA required a "black box warnings" to be included on labels of several anticancer drugs to indicate the increased risk of liver injury including some tyrosine kinase inhibitors.

D. 2. Kidney damage (nephrotoxicity)

Kidneys play an essential role in the urinary system. In addition, kidneys have other functions, which include the regulation of electrolytes, blood pressure, acid-base balance, re-absorption of glucose & amino acids, and the production of erythropoietin (a hormone that stimulates stem cells in the bone marrow to produce RBCs).

Kidneys filter out creatinine (a break-down product of phosphocreatine in muscle) from blood so that there is little re-absorption of creatinine. Therefore, higher blood levels of creatinine indicate kidney may be injured and has a problem filtering out creatinine. If the filtering function of the kidney is deficient, it usually indicates that significant damage has occurred to normal nephrons (the basic structural and functional unit of the kidney) and blood levels of creatinine will rise. Therefore, creatinine levels in blood and urine may be used to calculate the creatinine clearance as one measurement of renal function.

A more complete kidney function measurement also includes blood urea nitrogen (BUN). Urea is a substance secreted by the liver, and removed from the blood by the kidney. A greatly elevated BUN (>60 mg/dL) generally indicates a moderate-to-severe degree of renal failure.

Unlike conventional drugs, cisplatin is not really "detoxified" by the liver, thus not only can it damage liver, it can also reach the kidney in the same molecular structure and still not easily excreted by the kidneys. As a result, bold levels of creatinine and BUN are usually elevated by cisplatin treatment.

Nephrotoxicity is the major side effect of cisplatin, which evolves slowly but predictably after initial and repeated exposure of cisplatin [19-21]. Nephrotoxicity induced by cisplatin may be related to reactive oxygen species (ROS).

In addition, among the earliest reactions of the kidney is the activation of the mitogen-activated protein kinase (MAPK) cascade which is typical of the stress response. The inflammatory cascade is an important determinant of the degree of renal failure induced by cisplatin.

D.3. Cardiovascular toxicity (Cardiotoxicity)

Cardiotoxicity is associated with many anticancer drugs, particularly a very important class of anticancer drugs, anthracyclines. In fact, their use in cancer treatment is limited by a cumulative dose-dependent cardiotoxicity. Studies indicated that, at a cumulative dose of 550 mg/m^2, there was a significant increase in the development of heart failure; the use of smaller divided doses did decrease the chance of left ventricular ejection fraction. This cardiotoxicity induced by anthracyclines can cause irreversible heart failure and even fatality in some cases [22, 23].

In addition to anthracyclines, many other anticancer drugs may affect the cardiovascular system. In today's oncology, combination therapy is becoming more popular and this could also increase the chance of cardiotoxicity.

D.4. Mucositis

Mucositis refers to the inflammation on the mucous membrane in the digestive system. It is one of the side effects resulted from both chemotherapy and radiotherapy. It can occur anywhere along the gastrointestinal tract starting from the mouth. Mucositis in the mouth is often referred to as oral mucositis, which is a common side effect for patients treated by chemotherapy or radiotherapy [24]. Severe mucositis can lead to other complications including infections and difficulty to eat.

D.5. Peripheral neuropathy

Chemotherapy may affect the nervous system and causes tingling or decreased sensation in the hands, and/or feet; this side effect is often called peripheral neuropathy. Although peripheral neuropathy is often associated with diabetics, it is also a severe dose limiting toxicity associated with cancer chemotherapy [25, 26]. Detailed mechanism of peripheral neuropathy is not well understood, but it has been reported that it may be related to the damage of neuron DNA, dorsal root ganglion neurons, and some believe it could be mediated by oxidative stress [27, 28]. Chemotherapy-induced peripheral neuropathy can be reversible, therefore, reducing the dose of the medicine may eliminate the symptom. Sometimes, it may be necessary for the physician to change the medication.

D.6. Acral erythema

Acral erythema is a side effect of chemotherapy that can occur days to months after a patient begins treatment. Its symptoms include reddening, swelling, and desquamation of the skin, often of the palms and soles. Acral erythema has been reported in patients with prolonged treatment of cytotoxic anticancer drugs [29]. Usually its symptoms can disappear within weeks after discontinuation of the offending drug.

D.7. Ototoxicity

Ototoxicity is the toxicity on the function of the ear. It can be caused by a number of chemotherapy agents and antibiotics. The symptoms of ototoxicity include hearing loss, balance disorder, and ringing noise. Some anticancer drugs, such as cisplatin, can cause damage of hearing or even hearing loss [30].

PART II

RADIOTHERAPY AND ITS SIDE EFFECTS

Part II

Chapter 1
Introduction

Radiotherapy is also referred to as radiation therapy. Radiotherapy deals with the treatment of cancer and it should not be confused with radiology, which deals with the study and application of radiation as a diagnostic tool.

Radiotherapy is a method of treating cancer patients by using ionizing radiation, which consists of either subatomic particles or electromagnetic waves that are energetic enough to knock off electrons from atoms or molecules to ionize them. Ionizing radiation comes from two sources shown below:

(1) Radiation is emitted by radioactive substances including cobalt, cesium, radium, uranium, radon, and plutonium;
(2) Radiation is produced by medical devices such as an x-ray machine.

Examples of ionizing particles include α particles, β particles (electrons), protons, and neutrons; examples of electromagnetic waves include x-ray, gamma ray, and ultraviolet (UV).

Radiotherapy can be used to treat almost every type of solid tumor locally, including cancers of the breast, brain, lung, liver, cervix, pancreas, prostate, skin, stomach, and uterus. It can also be used to treat leukemia (cancer of blood-forming cells) and lymphoma (cancer of the lymphatic system, which is an important part of the immune system) [31]. In some cases, patients may receive a total body irradiation (TBI) for certain cancer, such as acute leukemia, as a pivotal part of bone marrow transplantation [32].

Radiotherapy has been used for both curative and adjuvant cancer treatment for many years. Adjuvant therapy refers to auxiliary/additional treatment for cancer; adjuvant radiotherapy is usually given after surgery or in conjunction with chemotherapy when a potential risk of relapse remains. Radiotherapy can also be given before surgery to reduce the mass of tumor to be removed.

Roughly speaking, 60% of all cancer patients receive radiotherapy, thus, radiotherapy might not be necessary for all cancer cases. For example, according to a scientific report in the *New England Journal of Medicine*, it was indicated that children with the most common form of leukemia can safely forego radiation therapy if they are treated with chemotherapy regimens tailored to their individual needs. The conclusion was based on a clinical trial involving 498 patients with acute lymphoblastic leukemia. Nearly 94 percent of the patients were still alive 5 years after treatment, a result that compares favorably with other treatment studies [33].

An overly simplified schematic representation of how radiation works is shown in Fig. 2.1. Basically, radiation therapy works by damaging the DNA of cancer cells and hampering these cells from replicating. The damage to cancer cells is caused by radiation, which directly or indirectly breaks up the DNA chain. The indirect damage comes from free radicals including hydroxyl radical ($HO^{\cdot}$), superoxide (O_2^{-}), hydrogen peroxide (H_2O_2), peroxyl ($ROO^{\cdot}$) and alkoxyl ($RO^{\cdot}$) radicals, collectively called reactive oxygen species (ROS) [34].

Indeed, most of the therapeutic effect (DNA damage in the fast-growing cancer cells) by radiation is resulted from the DNA damage in cancer cells. Unfortunately, some normal cells can also be damaged by radiation because they are also fast-growing. These fast growing normal cells include bone marrow cells, hair follicle cells, intestinal cells, and mucosal cells.

Fortunately, there is a very important cell function, DNA repair, by which a cell identifies and corrects damage to the DNA molecules. In fact, DNA repair process is constantly active as it responds to the continued damage in the DNA structure.

Because cancer cells are not completely differentiated, their DNA repair capabilities tend to be less complete than those of normal cells. Therefore, cancer cells tend to be more susceptible to radiation damage as compared to normal cells. On the other hand, cancer cells can still rely on residual DNA repair capabilities to survive damage induced by radiotherapy or chemotherapy; this is one of the reasons cancer is not easy to treat.

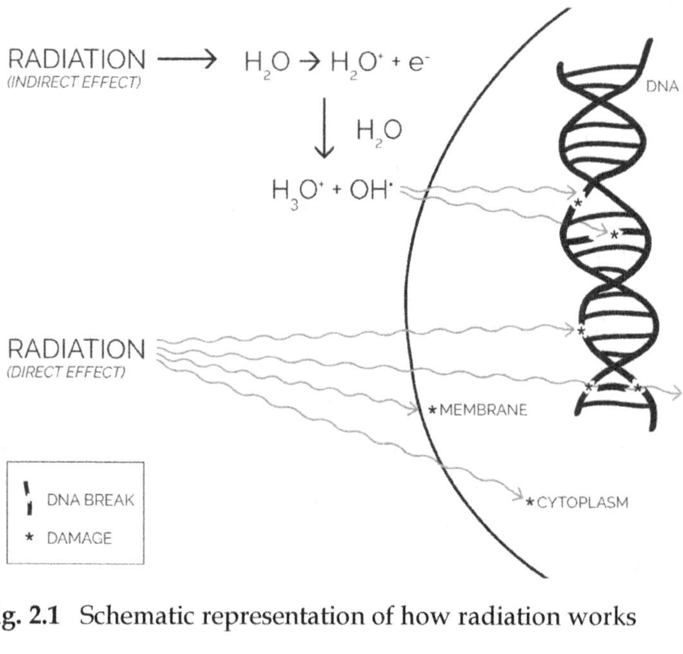

Fig. 2.1 Schematic representation of how radiation works

In addition to radiation-induced DNA damage to normal cells, there are additional undesirable biological effects on the body, including immunological responses (such as abnormal levels of some cytokines including TNF-α, TGF-β, IL-12, and PGE2). These side effects will be discussed in the later section.

Indeed, the goal of radiotherapy is to damage as many cancer cells as possible, while limiting the damage to nearby healthy tissues/organs as much as possible. Nevertheless, it is almost impossible to have zero damage on normal cells. That is why the treatment plan is to focus the radiation to tumor cells. Let's take a closer look at the side effects induced by radiotherapy in the following chapter.

Part II

Chapter 2
Side Effects of Radiotherapy

Although radiotherapy is a major and often effective modality in treating cancer patients, it can inflict serious damage on healthy tissues/organs. Because radiotherapy utilizes high energy radiation to kill cancer cells, injury to normal organ is almost inevitable. For example, the acute side effects of radiotherapy can occur on any surrounding normal tissues be it skin, brain or lung.

Fortunately, many of the side effects caused by radiotherapy are reversible, but some may not become apparent until months or even years later, such as secondary tumor formation. Fortunately, modern technologies in radiotherapy can significantly reduce the radiation exposure of normal tissues and reduce the side effects thereof.

In general, the side effects of radiotherapy can be summarized below:

1. Damage to the DNA of normal cells

The potential adverse effects of radiotherapy to the DNA of normal cells may result in tissue/organ damage and potentially secondary cancer. This is mainly due to (1) the direct exposure of normal cells to radiation, and (2) the indirect effect from the free radicals generated by radiation, which might result in the modification of certain genes generally referred to mutation. As a result, there is a possibility that certain cancer might be induced in the future, also known as secondary cancer, especially if high radiation dose is used often.

In recent years, significant technical improvements have been made to greatly reduce this type of potential side effect. Examples of the newer radiotherapy technologies are described in the following chapter.

2. Damage to cell renewal

The following side effects are under a class called cell renewal effects.

Damage to bone marrow

Bone marrow is the flexible tissue in the hollow interior of bones. It is responsible for the production of blood cells, including RBCs, WBCs, and platelets. If a person is exposed to a high dose of radiation for too long (especially on pelvis, chest and other bones), one can suffer bone marrow damage. As a result of radiotherapy, a patient may suffer from low WBC count (neutropenia or leukopenia), low RBC count (anemia), and/or low platelet count (thrombocytopenia). As explained in the previous chapter, neutropenia may increase a patient's chance of infection, anemia may cause fatigue, and thrombocytopenia can increase a patient's risk of bruising and bleeding.

On the other hand, radiation damage to bone marrow is not completely negative. For patients that require bone marrow transplantation, a high dose TBI becomes a positive tool in that unhealthy bone marrow cells are "wiped out" by high-dose radiation before a patient receives healthy bone marrow cells or cord blood cells.

Hair loss (alopecia)

Radiotherapy often causes hair loss on the body part being treated [35, 36]. However, hair loss from radiotherapy is usually limited to the area treated by the radiation. For patients who receive radiotherapy for brain cancer, the most pronounced side effect may be hair loss. In this case, radiation-induced hair loss could be permanent; this is different from chemotherapy on other part of the body wherein hair loss is more likely temporary.

Anemia

Depending on the dose and the time of treatment, radiotherapy may or may not cause anemia for cancer patients [37]. Many patients experience anemia at some point during treatment; in serious situations, it may cause aplastic anemia.

Aplastic anemia is when the hematopoietic stem cells in bone marrow are seriously damaged. As a result, all blood cells are not adequately produced, and could be fatal in some cases. One of the most important pioneers in radioactivity, Marie Curie, died from aplastic anemia which was believed to have resulted from her long-term exposure to the radioactive materials she studied.

Diarrhea

When the lining of mouth, stomach, and intestines are injured by radiotherapy, it may cause diarrhea. It is common to have diarrhea during or after radiotherapy treatment on the stomach or abdomen. The diarrhea can gradually get worse as the treatment goes on. Once the treatment ends, the diarrhea normally goes away gradually over a couple of weeks, but for some people it may take a little bit longer.

Skin damage

Radiotherapy can induce tanned (hyperpigmentation), irritated, peeling or burned skin. Because recent radiotherapy technology utilizes higher radiation doses, skin damage becomes a problem that happens more often. On the other hand, in the most recent radiosurgery, such as stereotactic radiotherapy, because radiation is applied from different direction, skin damage is significantly reduced.

3. Activation of immunologic responses

In addition to producing the direct DNA damage, radiation can also produce a lot of free radicals, including reactive oxygen species (ROS), and cause indirect damages to normal tissues/organs. The delicate immune system can also be disturbed by this oxidative stress, causing extraordinary number of immunological responses, including the abnormal release of many cytokines.

Cytokines are regulatory proteins released by cells of the immune system. It has been reported that a number of cytokines may be induced by radiation, including tumor necrosis factor alpha (TNF-α), IL-1, IL-4, IL-5, IL-6, IL-8, IL-10, and transforming growth factor beta (TGF-β) [38, 39]. The alteration of cytokine levels in blood, skin, or other tissues can cause a number of pharmacological responses.

4. Other side effects of radiotherapy

The medium- and long-term side effects of radiotherapy that one can see or feel depends on the tissues/organs that receive the radiation. With today's technological improvement, these side effects can be significantly reduced.

Because radiotherapy is being used for different types of cancers and every patient's situation is different, the side effects could vary from patient to patient. Some of the more common side effects from radiotherapy are described below:

Fatigue or tiredness

Fatigue is among the most common symptoms of radiotherapy [40]. Lack of energy and an overtired feeling are common symptoms. Depending on the type of cancer and the regimen of the radiation treatment, the degree of fatigue may be different. Usually, patients will gradually recover from the lack of energy several weeks after the completion of radiotherapy.

Dryness

Radiotherapy on patients with head and neck cancer often causes the release of free radicals in saliva glands, resulting in lack of saliva (abnormal dryness of the mouth or xerostomia). Radiation therapy may also cause dry eye (xerophthalmia). Similarly, vaginal mucosa may be dry after pelvic irradiation.

Fibrosis

Fibrosis is the development of excessive fibrous connective tissue in an organ or tissue. As a reactive response to the high radiation energy, irradiated tissues tend to be inflamed and gradually become less elastic. Overtime, certain irradiated area may gradually become fibrotic. This happens more often for lung cancer patients being treated with radiotherapy [36, 41].

Cancer (secondary malignancy)

Because radiation can alter certain DNA, it is also a potential cause of cancer, albeit a low probability. Some patients may develop secondary malignancies several years after receiving a course of radiation treatment. In most cases, this risk is greatly outweighed by the reduction of the primary cancer.

Additional information about the side effects of radiotherapy

Because a human body is a delicate and complicated biological system, and the inter-communication among different organs are even more complex, some side effects are inter-connected and may not be clearly separated from one another.

Cancer patients treated by high dose radiation to specific organ may suffer organ damage in addition to other side effects (hair loss, diarrhea, etc.). For example, patients with liver cancer, when treated by ionizing radiation, may suffer from liver damage and/or liver function reduction.

There are several reasons why organ damage occurs after radiation therapy. The obvious reason is that liver cells damaged by DNA will result in the damage of liver tissue through mechanisms such as programmed cell death (apoptosis). The other important reason is that certain cytokines, such as TNF-α, in liver and blood are often markedly induced by radiation therapy. As described before, abnormally elevated levels of TNF-α will kill the surrounding cells (cancer cells or normal cells), and will cause liver damage.

Our experimental results showed that radiation to mouse liver induced liver damage (reflected by elevated serum levels of both aspartate transaminase (AST) and alanine transaminase (ALT), and the liver damage is correlated with increased levels for serum TNF-α induced by irradiation [42]. Pre-treatment of mice with a small-molecule TNF-α modulator UTL-5d (Fig. 2.2) effectively reduced TNF-α secretion and resulted in marked reduction of ALT/AST levels [43]. This implies that TNF-α may play an important role in damaging the liver and UTL-5d compound may be a potential agent for radioprotection.

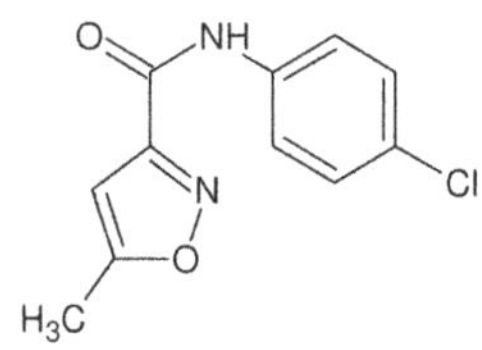

Fig. 2.2 Structure of UTL-5d

In order to reduce the side effects induced by radiotherapy, there are several general strategies being employed as described in the next chapter.

Part II

Chapter 3
Modern Technologies in Radiotherapy

Many approaches have been employed by the medical/pharmaceutical community to enhance the therapeutic effect of radiotherapy and/or to protect normal tissues or ameliorate tissue injury induced by radiotherapy. In general, there are four major approaches: (1) improvements/modifications on technical aspects of radiotherapy, (2) use of radiosensitizers to enhance the killing of radio-resistant cancer cells, (3) use of radioprotective agents to protect normal tissues and/or reduce the side effects, and (4) use of radiomitigators to speed up the recovery of normal tissues from injury induced by radiotherapy.

1. Improvements on the technical aspects of radiotherapy

Generally, radiation oncologists attempt to reduce the injury to normal tissues surrounding the target tumor by limiting (or blocking) the radiation field. In recent years, a number of significant technical developments and modifications in radiotherapy have been developed in order to increase the efficacy of radiotherapy and/or reduce radiotherapy-induced injury. These technology improvements include using multiple ports, higher radiation doses, interstitial brachytherapy, stereotactic radiosurgery, and TomoTherapy. In general, therapists use these approaches in attempts to increase the radiation dose to the tumor tissue while decreasing the dose to the surrounding normal tissue(s). The main goals are to enhance/maintain the therapeutic effect while minimizing the damages to normal tissues.

A representative example of these newer technologies is stereotactic radiotherapy. Stereotactic radiotherapy utilizes small fractions of high-dose radiation (as compared to conventional dose) from different angles to destroy tumor tissue. By giving multiple radiation doses from different angles, it can improve the radiotherapy outcome while minimizing potential side effects on normal tissues, especially the skin. Stereotactic radiotherapy is now frequently used to treat brain tumors and other types of solid tumors.

A specific form of stereotactic radiotherapy used for brain tumors is called stereotactic surgery, which utilizes high dose radiation (x-ray or gamma ray) to precisely focus on the specific area in the brain to destroy tumor tissues in the brain, very similar to surgery, thus often referred to as radiation surgery. In this treatment, the dose and the treatment area have to be precisely coordinated to ensure that normal tissues in the proximity suffer the least damage by the high dose radiation. It is also called fractionated stereotactic radiosurgery wherein the total dose of radiation is divided into several doses and delivered in several days.

Another popular new technology is called TomoTherapy. It is a computed tomograph (CT)-guided intensity modulated radiation therapy (IMRT) that delivers radiation to the tumor slice-by-slice, which is different from other forms of external beam radiation therapy in which the entire tumor volume is irradiated. TomoTherapy is essentially an image-guided radiation therapy (IGRT). The "beam on" time for TomoTherapy (3 to 5 minutes) is similar to normal radiotherapy. However, it does require additional daily CT to precisely locate the tumor because tumor size and position may shift due to the shrinkage over time. The daily CT will help the operator in directing the radiation beam to modify the treatment (adaptive radiotherapy). TomoTherapy has been used in many solid tumors including prostate cancer, lung cancer, as well as head and neck cancer.

In recent years, stereotactic radiotherapy is being combined with TomoTherapy. Right before the radiotherapy procedure, a patient is positioned on an examination table and a high-resolution CT scan is used to determine the precise location, size, and shape of the tumor. The data is then transferred to the computing system wherein an exact treatment plan is defined in real time. Once the patient is positioned, the radiation source is rotated and/or orientated by a computer to different positions/angles, and a predetermined dose of radiation is targeted at the tumor while sparing the healthy tissues. Commercially, several advanced radiotherapy systems utilizing similar principle have been developed and used. Examples of modern radiotherapy systems include Discovery CT590 RT and Optima CT580 made by GE Healthcare, and CyberKnife® made by Accuray. The most recent comprehensive radiosurgery system, Edge™, is made by Varian Medical Systems. Edge™ is just entering the market and there are only few units installed in the United States. It is a dedicated system for performing advanced radiosurgery using new real-time tumor tracking technology.

2. Improvement of radiotherapy by using radiosensitizers

A radiosensitizer (also referred to as a radiosensitizing agent) is a pharmaceutical agent that makes tumor cells more sensitive to radiation, thus, increasing the efficiency of radiotherapy. For example, oxygen is known to increase tumor sensitivity to radiation by forming free radicals that can damage DNA.

As a solid tumor continues to grow, some cancer cells outgrow their blood supply and become sufficiently removed from their capillaries to become hypoxic (lack of oxygen). As a result, cancer cells survive by another mechanism called glycolysis. Glycolysis is a process in which glucose is converted into pyruvate by an oxygen-independent metabolic pathway; the process produces energy which is then stored in the form of ATP (adenosine triphosphate) for anaerobic cancer cells. In the presence of oxygen, ATP is formed through oxidative phosphorylation.

The process of glycolysis further results in excess lactic acid, which makes solid tumor more acidic (pH can be as low as 6.2) [44-46]. In general, cancer cells under acidic condition (mainly due to hypoxia) are more resistant to killing by radiotherapy and some cancer drugs.

Currently, there are a number of agents being investigated as potential radiosensitizers. Some examples of radiosensitizers are shown in Part III.

3. Improvement of radiotherapy by using radioprotectors

Radioprotection is another important approach in radiotherapy. Basically, a therapeutic agent, radioprotector, is administered to a patient before the patient is treated by radiotherapy. In other words, radioprotectors protect normal tissues against radiation-induced side effects. Ideally, radioprotectors produce or stimulate certain biological reactions/products which then counteract the adverse effects caused by radiation.

Unfortunately, there has not been any significantly improved radioprotector since the discovery and approval of a popular broad-spectrum radioprotective drug, amifostine (Fig. 2.3), by the FDA. Interestingly, amifostine was originally approved to reduce the cumulative renal toxicity associated with repeated administration of cisplatin in patients with advanced ovarian cancer and non-small cell lung cancer [47]. However, while nephroprotection was observed, amifostine also formed chelation with cisplatin and reduced its anticancer activity [48]. Therefore, its indication for chemoprotection in cisplatin treatment in non-small cell lung cancer was withdrawn in 2005.

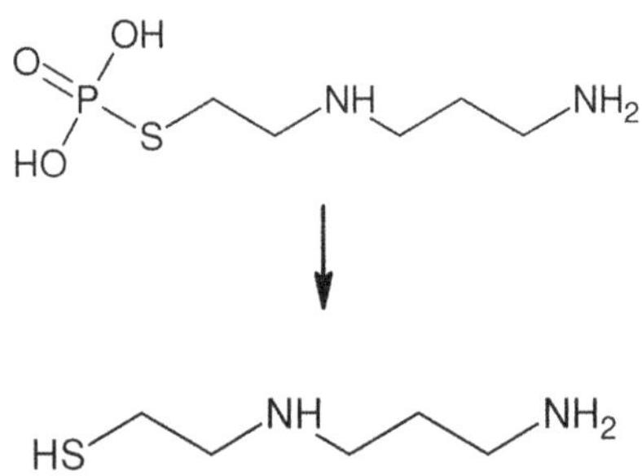

Fig. 2.3. Structure of amifostine (WR2721) and its metabolite

More details about amifostine and other radioprotectors are described in Part III.

4. Improvement of radiotherapy by using radiomitigators

Different from radioprotectors, which need to be administered either before or during radiotherapy, radiomitigators can be administered after radiation treatment. Radiomitigation can also be very useful in term of national defense with the increasing global threat of terrorists, such as potential radiological attack by dirty bombs.

Currently, there is not an FDA approved radiomitigator. However, there are a number of drug candidates under investigation as radiomitigator.

Part III

Management of the Side Effects

Part III

Chapter 1
How to Cope with the Side Effects

The major side effects of chemotherapy and radiotherapy include tissue/organ injury which occasionally may be fatal. In addition, these side effects may prevent doctors from delivering the prescribed dose of therapy according to the predetermined schedule in the cancer treatment plan. It is important to understand that the expected outcome from therapy is based on delivering treatment at the dose and schedule according to the well-designed and predetermined treatment plan. In other words, not only are the side effects of chemotherapy or radiotherapy undesirable, they may also limit a patient's ability to accept the complete therapeutic regimen (both dose and schedule) to achieve the best outcome.

Three Types of Side Effects

In general, side effects from chemotherapy and radiotherapy can be divided into three types. The first is a set of physiological responses, which are acute and the least concerning, including appetite loss, nausea and vomiting, weakness and fatigue, as well as flu-like symptoms; all of these can be managed by specific actions.

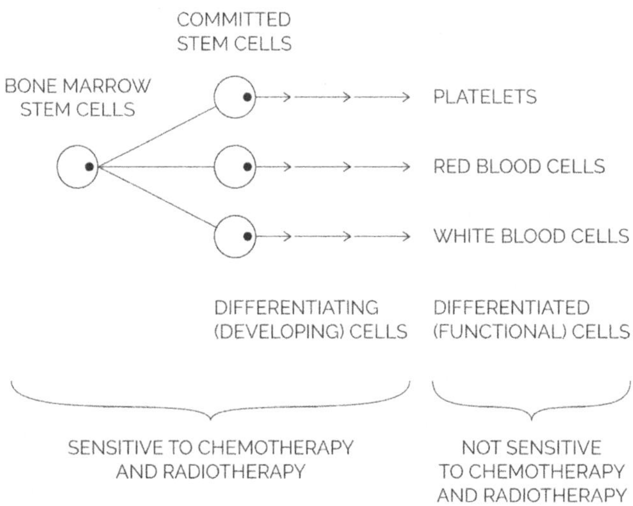

Fig. 3.1 Common cell types in a cell renewal system

The second type of side effects is more severe and may damage cells in renewal tissues. There are a number of such cell renewal tissues in the body encompassing the hematopoietic system, the gastrointestinal tract, skin, hair, gonads, and cornea. Fig. 3.1 shows the common cell types in a cell types in a cell renewal system. The first and basic cell type is the stem cell - destroy all of these and the tissue gradually disappears. These cells are especially sensitive to both radiation and chemotherapeutic agents, and more so when they are proliferating. If over 99.9% of the stem cells are destroyed, the renewal tissue can't recover by itself and the host will die. If stem cells are not totally destroyed, stem cell transplant and some protective agents can ameliorate some of the damage. The "committed" stem cells and differentiating cells (which may require proliferation) are also sensitive to chemotherapy and radiotherapy since they are proliferating just like the cancer cells. Finally, the differentiated cells such as platelets and red blood cells are resistant to chemotherapy. However, since the differentiated cells have a specific half-life, they are "used up" and may not be replaced as the stem cells and differentiating precursor cells have been destroyed.

The third type of side effects is direct damage to cells in major organs such as liver, heart, and brain. These are non-proliferating cells, and like the differentiated cells discussed above and relatively insensitive. However, some drugs do have selective toxicity to specific organs such as anthracyclines and cardiac toxicity.

Coping with the Side Effects

Before, during, and after radiotherapy or chemotherapy, there are some prescription drugs and natural products that can be used to reduce specific side effects induced by the therapy. In addition, there are also some activities that patients can do to reduce the discomfort from cancer therapies. Let's look at them according to the types of side effects.

Dealing with Type 1 Side Effects

In general, patients should first consult their doctors to find out what medications are expected to work best for their situations. Other activities are described under individual side effects below.

Fever and Infection

Chemotherapy often lowers WBC count in the blood resulting in neutropenia. Because white blood cells are very important in fighting against infection, when a patient's WBC count is lower than normal level, the patient is more susceptible to infection.

To reduce the chance of infection, patients should avoid going to public areas, such as theater and shopping mall. Also, avoid close contact with anyone who is ill. Wash hands frequently and when washing hands, use appropriate hand soap especially when the hands are visibly soiled. In recent years, alcohol-based hand sanitizer is becoming popular and it can be very effective in reducing the number of germs. In addition, do not eat raw food including fish, beef, oyster, and food containing raw egg. Thoroughly wash the vegetables and fruits for salad. Wash the cutting board especially after cutting raw meat, poultry, or fish. A patients should call his doctors if the body temperature reaches 101 °F (or 38.3 °C) or higher.

Flu-Like Symptoms

After a chemotherapy treatment, some people may experience flu-like symptoms such as pains and muscle aches. If not serious, patients can take over-the-counter medications such as Tylenol or Advil to relieve the pain. However, if the symptom persists, patients should contact their doctors for prescription drugs. In many states and Washington D.C., marijuana may be prescribed by physicians for pain management.

Nausea and Vomiting

For the management of nausea, patients should try not to skip meals. An empty stomach can worsen the symptom. Eat small and light meals. Drink lots of fluids, such as water, tea, and juice. Acupressure is being used by some people to reduce nausea basically by pressing a specific acupressure point near the base of the arm. There are special wristbands available for such an acupressure; the wristband usually has a knob to apply pressure to the correct spot. Further, there are some anti-nausea over-the-counter drugs available to prevent and treat nausea and vomiting due to chemotherapy.

Fatigue

Fatigue is a common symptom for people treated for cancer. For some, the fatigue may last a long time after treatment, and it can discourage them from physical activity. However, proper exercise can actually help patients reduce fatigue. Physical therapy can also help patients maintain strength in the muscles, which could help fight the sense of depression that sometimes comes with physical fatigue.

Most cancer drugs can reduce the RBC count and may cause anemia, which can make patients feel tired easily. Most people have to make some adjustment in work and family responsibilities. For anemia patients, it's a good idea to plan their major activities and try to maintain their regular daily activities as much as possible. The fatigue should gradually go away after the patients recover from chemotherapy.

Doing regular exercise such as yoga, taichi, and walking may be helpful in relaxation. If necessary, physicians can prescribe iron supplement or other newer drugs to boost the blood cells. These include Epogen (by Amgen), Neupogen (by Amgen), and Leukine (by Berlex), which function as colony-stimulating factors because of their ability to stimulate the formation of blood cells in the bone marrow.

Appetite and Taste Changes

During chemotherapy, one may lose appetite and becomes more sensitive to odors. As a patient becomes healthier, his/her appetite will improve. Again, one should try to stay with healthful food (low fat, low calorie, and high fiber) and drink a lot of liquid. Also, it's a good idea to avoid spicy food. For some people, eating too much spicy food can result in diarrhea.

With a loss of appetite, important nutrients may be lacking. Suitable dietary supplement products may be used to supply needed nutrition.

Menopause

Depending on a woman's age and the type of chemotherapy administered, chemotherapy may temporarily stop the periods. As a result, one may experience symptoms such as hot flashes, decreased libido, vaginal dryness, mood changes, and sleeping disturbances. If a patient experiences any of these symptoms, she should call her doctor or nurse about suitable treatment for the symptoms. Meanwhile, consider eating soy products (such as soy milk and tofu) which may reduce hot flashes. Wear light cotton pajamas to help prevent overheating when sleeping. Use vaginal moisturizers or other water based lubricants as needed.

<u>Dealing with Type 2 Side Effects</u>

Anemia (lower levels of red blood cells)

As mentioned before, both chemotherapy and radiotherapy can lower RBC count and result in anemia. Generally, the RBC count for a healthy adult is about $5 \times 10^6/\mu L$. Because chemotherapy or radiotherapy could damage bone marrow, which produces RBCs, RBCs are not replaced at the same rate as they are being destroyed. As a result, the RBC count becomes much lower than that in a healthy person, a situation referred to as anemia.

If patients become anemic due to cancer therapies, they will have reduced blood levels of RBCs, which carry oxygen from the lungs to the body's muscles and other organs. As a result, patients can become tired more easily; patients may also feel dizzy or light-headed when they quickly stand up from a crouching position. Although a normal healthy body can replenish lower levels of RBCs, it does take time. In addition, multiple treatments of chemotherapy or radiotherapy can hamper the reproduction process of RBCs. During the cancer therapy, when the RBC count is significantly below acceptable levels, it could cause a physician to temporarily suspend the chemotherapy or radiotherapy. In more serious conditions, blood transfusion or treatment by a specific bone marrow stimulating drug(s) may become necessary. Some of these drugs will be briefly described in the following sections.

Otherwise, what can a patient do? First, a patient should watch the diet. For example, eat nutritious food including liver, red meat, and green vegetables (such as spinach) to supplement iron in blood, which is important for maintaining RBC levels. Most of the body's iron is bound to hemoglobin in RBCs. When red blood cells die, the released iron is carried by transferrin (iron-binding plasma glycoprotein) to the bone marrow, to make new RBCs.

Iron pills can be very useful for treating anemia, but they should be taken under the prescription of the physician because of the potential side effects. When too much iron is taken, it could slowly accumulate and could become toxic in different organs, including the liver and heart.

In addition to iron supplement, patients should take some other dietary supplements, such as vitamin C, which is important for iron absorption, and vitamin B_{12}, which is important in maintaining normal RBC formation. Furthermore, patients should rest often to conserve energy and not to conduct any vigorous exercises.

Leukopenia (low WBC count)

There are reports about a natural chemical, hesperidin, which may be helpful in increasing WBC count. Hesperidin is abundant in the peels of citrus fruit, including orange, lemon, and grapefruit. One can be creative in using the peel by blending the peel in juice or in food for easier consumption.

Thrombocytopenia (low platelet count)

Although it does not happen for all chemotherapeutic agents, thrombocytopenia does happen to many patients. Thrombocytopenia is not uncommon for acute leukemia patients treated with radiotherapy; under serious condition, it could even be life-threatening. Similar to anemia and neutropenia, thrombocytopenia is usually temporary. When the platelet count is low (between 15,000 and 30,000 per µL), unexplained bruises may happen. When the platelet count is even lower (under 10,000 per µL), a patient may need a platelet transfusion.

As a patient, one must be careful in daily activities and try to avoid any injury because the bleeding might take much longer than normal to stop. In addition, patients need to be careful in conducting any physical activities to avoid cutting/bleeding.

Hair Loss

Not all chemotherapy drugs cause hair loss. Some people may lose all the hair on their heads and other body hair, while others may not lose much hair. For many people, hair loss may be a difficult aspect to deal with. In this case, using a wig to cover the head could make a patient feel more comfortable. Several weeks after the therapy, the hair should begin to grow back; the new hair may be slightly different in color and texture. Usually, the new hair will be thin and soft, like baby's hair, but will often return to its original color and texture after several months.

Diarrhea or Constipation

Crypt cells in the intestine are likely to be attacked by chemotherapy drugs and may result in diarrhea. Some chemotherapeutic drugs may cause constipation. Patients with constipation should consider taking a stool softener, such as Colace® or Senekot®. Also, drinking a lot of liquid may help prevent constipation. If a patient has several watery stools or constipation, the patient may want to call his/her doctor or nurse.

Mouth Sores

Because epithelial cells in the mouth and esophagus are fast dividing, they are easily damaged by chemotherapy and radiotherapy. As a result, mouth sores occur often with chemotherapy and radiotherapy. For a patient to assuage that, rinse the mouth often with salty water. If a patient develops painful mouth sores and the situation is not relieved, one should call his/her physician or health practitioner and a special mouth rinse may be prescribed for the patient. If a commercial mouth wash is used, avoid those that may be irritating to the mouth.

Dealing with Type 3 Side Effects

Liver and Kidney Injuries

Liver is the most critical organ in the body's metabolism and detoxification. In addition, it is a vital organ capable of regenerating some of the lost tissue. For liver injury, some drugs may be helpful in protecting liver. For example, amifostine (Trade name Ethyol®) is indicated to reduce the cumulative renal toxicity associated with repeated administration of cisplatin in patients with advanced ovarian cancer. Other than amifostine, some natural products, such as milk thistle, may be useful in protecting or improving liver health.

Kidney toxicity is also referred to as nephrotoxicity, which is one of the most common toxicity resulted from chemotherapy. There are two important indicators of nephrotoxicity: BUN and creatinine. When kidney is damaged, blood levels of BUN and creatinine are usually elevated. Because urea is a waste form of nitrogen in the body, elevated BUN levels indicate that there is extra nitrogenous waste in the blood stream not properly filtered off by the kidney. As to creatinine, it is produced from creatine, an important molecule in energy production of muscle. Creatine is constantly converted into creatinine, which is constantly filtered off by the kidneys and disposed of in the urine.

In other words, kidney is critical in filtering off both BUN and creatinine to remove waste from the body. Therefore, when kidney function is affected by chemotherapy, blood levels of both BUN and creatinine are elevated.

Peripheral Neuropathy

Neuropathy means disease or dysfunction of the nerves, which can be a side effect of chemotherapy. For peripheral neuropathy, the most commonly affected areas are the tips of fingers and toes. Common symptoms of peripheral neuropathy include tingling, burning, numbness/pain in the affected areas, and loss of balance, although sometimes other areas may also be affected. Massage one's hands and feet may help improve blood circulation and temporarily relieve the pain. For a patient with peripheral neuropathy, he/she should see a doctor about any neuropathy symptoms to prevent further nerve damage.

Part III

Chapter 2
Drugs and Potential Drugs as Chemoprotectors

In this chapter, we will discuss drugs and potential drugs (often referred to as agents) that may be used before chemotherapy to reduce the side effects. These drugs/agents are often referred to as chemoprotectors. Examples of chemoprotectors are listed below based on the individual targets to be protected.

1. Drugs to Increase White Blood Cell Count

Filgrastim

Filgrastim is a 175-amino acid human granulocyte colony-stimulating factor (G-CSF) manufactured by recombinant DNA technology in *E Coli*. It is a glyprotein naturally produced by a number of different tissues to stimulate the bone marrow cells to produce granulocytes and release them into the blood.

Filgrastim is used to stimulate the proliferation and differentiation of neutrophils, which are the most abundant type of WBCs. It is marketed as Neupogen® by Amgen, as Religrast® by Reliance Biopharmaceuticals, and as Shilgrast™ by Raichem Lifesciences.

Sargramostim

Sargramostim is a WBC stimulant sold as Leukine® by Genzyme. It is a human granulocyte macrophage colony stimulating factor (GM-CSF) produced by recombinant DNA technology in a yeast, *S. cerevisiae*. Sargramostim is a glycoprotein having 127 amino acids. Its amino acid sequence is different from the natural human GM-CSF by a substitution of leucine at position 23. It was originally approved by the FDA in March, 1991 for use following autologous bone marrow transplantation (BMT) and was approved in December 1991 to prevent death following BMT engraftment delay or failure. In September of 1995, it was subsequently licensed to shorten the time for neutrophil recovery to prevent early death from life-threatening infections following chemotherapy for older patients suffering from acute myelogenous leukemia (AML).

2. Drugs to Increase Red Blood Cell Count

For cancer patients receiving chemotherapy, sometimes anemia may be incurred. When it happens, there are several treatment options. One option is to receive red blood cell transfusion. Another option is to be treated by erythropoiesis-stimulating agents, such as epoetin alfa [49, 50].

Epoetin alfa

Epoetin alfa (sold by Amgen as EPOGEN®) is a glycoprotein hormone which elevates or maintains the RBC level and is used to treat anemia. For patients with chronic renal failure or with damaged kidneys by chemotherapy, the main reason of anemia is their impaired production of erythropoietin (EPO) [51, 52].

EPO stimulates the production of RBC (erythropoiesis) by enhancing the division and differentiation of committed erythroid progenitors in the bone marrow. EPO is produced by the peritubular capillary endothelial cells in the kidney and liver. Levels of EPO are normally regulated by tissue oxygenation. Low oxygen (hypoxia) and anemia generally increase the production of erythropoietin to stimulate erythropoiesis.

Epoetin alfa is a glycoprotein containing 165 amino acids made by recombinant DNA technology. It contains the identical amino acid sequence of natural EPO and has the same biological effects as endogeous EPO. EPOGEN® is indicated to elevate or maintain the RBC level and is used to treat anemia for patients with chronic renal failure.

3. Drugs to Increase Platelet Count

Platelets (or thrombocytes) are small anuclear cell fragments (2-3 μm in diameter) differentiated from and much smaller than megakaryocytes (~19 μm in diameter for adults) [53]. They are large bone marrow cells with a lobulated nucleus responsible for the production of blood platelets. The condition of low platelet count is called thrombocytopenia.

Both liver and kidney produce a hormone called thrombopoietin (TPO or platelet growth factor), which stimulates the production of megakaryocytes. Therefore, damage to liver and/or kidney from chemotherapy or radiotherapy can result in reduced production of TPO, thus reduced platelet count.

Platelets play a critical role in hemostasis (stoppage of bleeding). Platelets also secrete platelet-derived growth factor (PDGF), which plays a role in cell proliferation, wound healing, and cell migration.

The average lifespan of a platelet is about 10 days. Under normal condition, as the platelets gradually die off, similar numbers of new platelets are produced by megakaryocytes in bone marrow to maintain the normal levels in the circulatory system.

Radiotherapy and chemotherapy can damage platelets and result in low platelet count. Because platelets play an important role in blood clotting, patients with thrombocytopenia should avoid any activities that can result in bleeding. By carefully monitoring platelet count and taking necessary action, it is possible to minimize bone marrow failure.

Normal platelet count ranges from 150,000 - 400,000 per µL of blood; chemotherapy or radiotherapy may drop the platelet count significantly. When the platelet count drops below 50,000 per µL, a patient is considered to be at increased risk for bleeding. For someone with a platelet count below 10,000 per µL, a transfusion of platelets may be necessary.

Until recently, transfusion was the only treatment available for patients with low platelet levels. While there are several drugs approved for idiopathic thrombocytopenic purpura (ITP), which is a condition of having a low platelet count with no known cause, none of these drugs are approved for chemoprotection or radioprotection.

Romiplostim

In 2008, the FDA approved a protein analog of TPO called romiplostim (sold by Amgen as Nplate®) as an orphan drug. Romiplostin is indicated as a potential treatment for chronic ITP [54]. Romiplostim treatment is administered by a variable dosage determined by analysis of the patient's platelet count at the time of treatment.

Romiplostim is produced by recombinant DNA technology in *E. Coli*. It promotes the megakaryocyte colony-forming cells and TPO-dependent cells; *in vitro* studies in human and murine megakaryocytes showed that romiplostim binds to TPO receptor [55, 56].

Eltrombopag

Another drug approved by the FDA in 2008 is eltrombopag (Fig. 3.2). Eltrobopag is the active ingredient of Promacta® (sold by GlaxoSmithKline). It is the first non-peptide small-molecule mimetic TPO approved by the FDA, thus a significant milestone in thrombocytopenia therapy.

Eltrombopag is a small-molecule drug for the short term (6 week) treatment of ITP in adults. Eltrombopag is a TPO receptor agonist indicated for the treatment of thrombocytopenia in patients with chronic immune/ITP who have had an insufficient response to corticosteroids, immunoglobulins, or splenectomy. It should be used only in patients with ITP whose degree of thrombocytopenia and clinical condition increase the risk of bleeding.

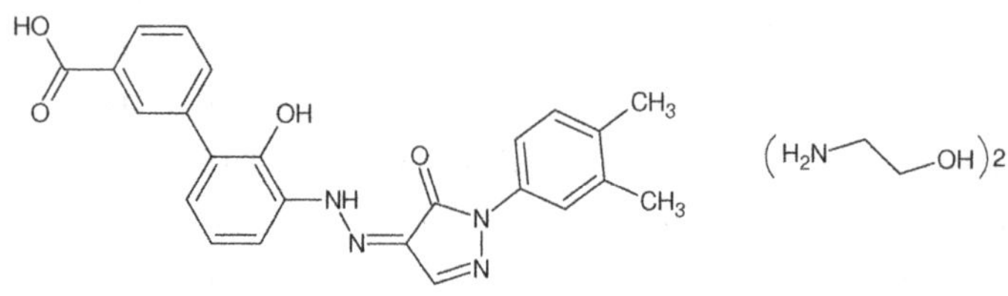

Fig. 3.2 Structure of eltrombopag olamine

4. Drugs for Bone Marrow Protection

Plerixafor

Plerixafor (Figure 3.3) is a small-molecule CXCR4 chemokine receptor antagonist. Its dosage form is a colorless to pale yellow isotonic subcutaneous injection (sold as Mozobil™ by Genzyme). Plerixafor is a strong inducer to mobilize hematopoietic stem cells from bone marrow to the bloodstream as peripheral blood stem cells [57]. Peripheral stem cells are like immature blood cells in the circulating blood. Increased peripheral stem cells help produce healthy blood cells in patients.

Mozobil was approved by the FDA in December, 2008, indicated in combination with G-CSF to mobilize hematopoietic stem cells to the peripheral blood (i.e., in blood circulation) for subsequent autologous transplantation in patients with non-Hodgkin's lymphoma and multiple myeloma. Autologous transplantation uses an individual's own blood as opposed to allogenic transplantation in which someone else's blood is used.

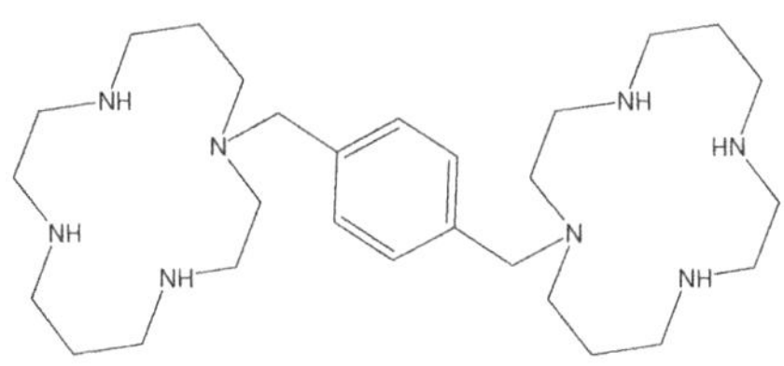

Fig. 3.3 Structure of plerixafor

Adverse reactions of Mozobil include diarrhea, nausea, headache, dizziness, and vomiting. Just like many other drugs, there are a number of precautions for using Mozibil. These include: (1) Mozobil may mobilize leukemic cells and should not be used in leukemia patients. (2) Mozobil may release tumor cells from bone marrow, and (3) Mozobil may increase circulating leukocytes (leukocytosis) and decrease platelet counts.

Pegfilgrastim (sold as Neulasta® by Amgen)

Suppression of WBCs is called myelosuppression. A disorder with low number of neutrophils (the most abundant type of WBCs) is called neutropenia, which can put patients at risk of severe infection. Therefore, chemotherapy regimen is usually halted until WBC count goes back to normal levels [58, 59].

Pegfilgrastim is a man-made protein that may be categorized as a colony-stimulating factor. It can stimulate the bone marrow and promotes the growth of neutrophils.

Neulasta® is a prescription drug to boost the production of WBCs to reduce a patient's risk of infection. As a result, the chemotherapy regimen can be maintained and increase the chance of a successful chemotherapy [60]. Pegfilgrastim is given to cancer patients 24 hours after chemotherapy is administered.

Similar to many other prescription drugs, Neulasta® is associated with certain side effects. These include abdominal pain, shortness of breath, trouble breathing, or shortness of breath.

5. Drugs for Mucositis

Palifermin

Keratinocyte growth factor (KGF) is a growth factor present in wound healing process. Palifermin is a protein similar to endogenous human KGF, but with the first 23 N-terminal amino acids removed to improve the protein stability. It is supplied as a sterile lyophilized powder for *i.v.* injection after reconstitution with sterile water.

Palifermin was approved by the FDA in 2004 and sold by Amgen as Kepivance™. It is indicated to decrease severe mucositis for patients treated with TBI; in order to prepare patients for stem-cell transplant, TBI is often required.

In addition to the approved drugs, some natural products have been reported to be helpful in stimulating RBCs, WBCs, and platelets. Examples include cordyceps sinensis, and hesperidin. These natural products will be discussed later.

6. Drugs for Cardioprotection

Zinecard® (dexrazoxane, Figure 3.4) is a cardioprotective agent for use to protect the heart from the cumulative side effect of anthracyclines (for example, doxorubicin).

According to the FDA approved label, it is a sterile, pyrogen-free lyophilized powder intended for intravenous injection. Chemically, dexrazoxane is a white crystalline powder which melts at 191 to 197 °C. It is sparingly soluble in water and 0.1 N HCl, slightly soluble in ethanol and methanol and practically insoluble in nonpolar organic solvents. Dexrazoxane has an octanol/water partition coefficient of 0.025 and degrades rapidly above a pH of 7.0. The structural formula is shown as follows:

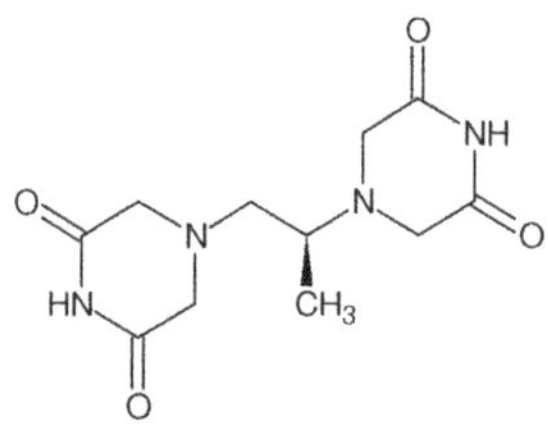

Fig. 3.4 Structure of dexrazoxane

Dexrazoxane is a cyclic derivative of EDTA that readily penetrates cell membranes. It is converted intracellularly to a ring-opened chelating agent that interferes with iron-mediated free radical generation thought to be responsible, in part, for anthracycline induced cardiomyopathy.

7. Drugs for Peripheral Neuropathy

As mentioned in the previous sections, peripheral neuropathy is quite common following chemotherapy or radiotherapy. However, the underlying mechanism behind this adverse neural effect is not clear. Currently, there is no FDA approved drug specifically for treating peripheral neuropathy although some drugs are being used off label to relieve the symptoms. These include pain relievers (such as oxycodone and tramadol) and antidepressants (such as doxepin). Therefore, it is important to investigate the mechanism by which chemotherapy/radiotherapy produce this neurotoxicity.

One of the research areas is DNA damage/repair with regard to peripheral neuropathy. It is known that chemotherapy and radiotherapy can result in DNA damage (e.g., alkylation and oxidation) and DNA repair mechanisms play a critical role in reducing the toxicity to the neuronal tissue. Because endonucleases play a role in DNA repair, it can be important in mechanistic studies and intervention in peripheral neuropathy.

Apurinic endonuclease (APE) is an enzyme that is involved in the DNA base excision (BER) repair pathway. When de-purination occurs, the APE recognizes the lesion, cuts DNA at this site to allow for DNA repair to continue[61, 62]. In Eukaryotes, there is only one AP endonuclease, APE 1 (also known as HAP 1 or Ref-1), which acts as an apurinic (AP) lyase by hydrolyzing the phosphodiester backbone at the 5′ end of an AP site [63]. Therefore, APE1 is an important target for mechanistic studies and intervention for peripheral neuropathy.

8. Broad-Spectrum Chemoprotectors

<u>Amifostine</u>

There are drugs/agents that protect more than just one specific biological target or even the total body. One of the most known broad-spectrum protectors is amifostine.

Aminothiols are a class of compounds most studied in radioprotection. A number of aminothiols have been studied; the most known radioprotector is amifostine (Fig. 2.3). Because amifostine is the most prescribed broad-spectrum radioprotector, it will be described in more details below.

Amifostine was first approved as a chemoprotector to reduce the cumulative renal toxicity associated with repeated administration of cisplatin in patients with advanced ovarian cancer and NSCLC in 1996.

After many years of research, amifostine remains the only FDA approved drug in the US as a radioprotector to reduce xerostomia (abnormal dryness of the mouth) associated with radiotherapy of head-and-neck cancer, albeit it has several significant drawbacks (Ethyol package insert, 2008) [64].

Amifostine is a phosphorylated aminosulfhydryl compound; its chemical name is S-2-(3-Aminopropylamino)ethylphosphorothioic acid. In the early stage, amifostine was referred to as WR-2721. The pharmaceutical end product of amifostine is Ethyol®.

Although amifostine is an FDA approved radioprotector/chemoprotector, there are significant limitations associated with amifostine including:

As a chemoprotector, the major limitation of amifostine is that simultaneous administration of a sulfur-containing chemoprotector with cisplatin has shown rapid chemical quenching of cisplatin by the sulfur anion or the sulfhydryl moiety to form an inactive platinum–thiol conjugate species [48]. Animal data suggest that amifostine may have tumor-protective effect and thus amifostine is not recommended with curative chemotherapy outside of a clinical trial [65, 66]. In addition, both amifostine and its analog, WR-1065, have very short half-lives and are rapidly cleared from the plasma (95% and 50% of the peak concentration within 1 h, respectively) [67].

Amifostine is usually administered by a 15 minute infusion 30 minutes before cisplatin chemotherapy, which is inconvenient [Ethyol, 2010].

Therefore, there is ample room for significant improvement on amifostine as a broad-spectrum radioprotector. An ideal chemoprotector should be orally administratable, with a longer half-life as compared to amifostine, and with little or no toxicity by itself.

UTL-5g

UTL-5g is a small-molecule immunomodulator been investigated by our group for a few years. It was first developed as a potential anti-arthritic agent and gradually evolved to become a broad-spectrum chemoprotective and/or radioprotective agent. Early studies of its chemoprotective effects were conducted mainly on mice treated by cisplatin.

Since cisplatin is a very important anticancer drug and also the foundation of all platinum drugs, it makes sense to use cisplatin as a representative example to describe several broad-spectrum chemoprotectors, amifostine and UTL-5 series of compounds. Let's focus on three important side effects: nephrotoxicity, hepatotoxicity, and myelosuppression.

The clinical use of cisplatin is mainly limited by its side effects such as nephrotoxicity which evolves slowly but predictably after initial and repeated exposure [20]. Among the earliest reactions of the kidney is the activation of the mitogen-activated protein (MAP) kinase cascade and molecular responses typical of the stress response. The MAP kinase cascade and inflammatory responses are important determinants of the degree of renal failure induced by cisplatin. It was reported that manipulation of these responses may be exploited to reduce the toxicity of cisplatin clinically [67]. One important response is related to TNF-α, which is a cytokine playing important roles in cancer and immune-mediated diseases. Since the production of TNF-α is coordinated with the multiple MAP kinase pathways, we believe that modulation of TNF-α is a legitimate biological target of chemoprotection and have conducted a number of animal studies for this investigation.

Recently, we investigated a small-molecule TNF-α modulator, UTL-5g (Fig. 3.5), as a potential chemoprotective agent. In our mouse studies, the results showed that UTL-5g lowered the elevated levels of BUN and creatinine induced by cisplatin. What does this result mean?

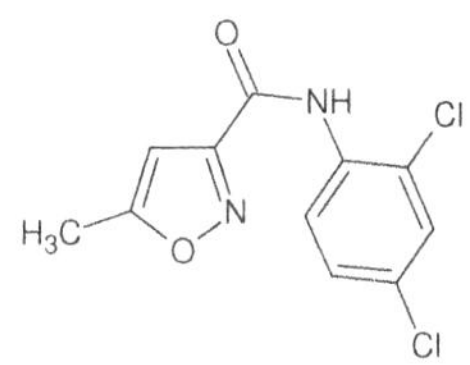

Fig. 3.5 Structure of UTL-5g

Normally, the liver produces urea as a waste product from protein. Normal human adult blood maintains 7 to 21 mg of urea nitrogen per 100 mL. Any extra urea in the blood is removed by the kidney. If BUN is elevated over normal range, it indicates that kidney function may be impaired. As to creatinine, it is produced at a fairly constant rate by the body and removed from the blood mostly by the kidney, thus blood creatinine is another important indicator of renal health. Therefore, our results indicate that UTL-5g could reduce the nephrotoxicity induced by cisplatin [42].

Another significant side effect of cisplatin is hepatotoxicity (liver toxicity), which is the dose-limiting toxicity of cisplatin. Hepatotoxicity occurs when cisplatin is administered at high doses [68-70]. Because liver is a very important organ, this problem is critical and must be addressed. Again, UTL-5g showed significant protection of the liver as indicated by its effect in lowering elevated levels of AST and ALT induced by cisplatin [42].

Yet another side effect of cisplatin is myelosuppression, which occurs in 25 to 30 percent of patients who undergo treatments with cisplatin [71, 72]. Myelosuppression relates to a reduction of activity in the bone marrow, in particular, toxicity to the blood forming elements. Generally, platelets and WBCs (or leukocytes) associated with myelosuppression are at the lowest levels about three weeks after treatment and return to normal two to three weeks thereafter. The loss of platelets and the loss of leukocytes are more pronounced when higher doses of cisplatin are given. In addition to the loss of platelets and WBCs, the use of cisplatin could also cause a decrease in the number of RBCs especially for a long-term treatment. We have published several papers related to the reduction of cisplatin toxicity and will talk more about UTL-5g in the later sections.

In summary, UTL-5g protects mice treated by cisplatin in different aspects. It protects kidney, liver, and bone marrow (as indicated by its protection on WBCs and platelets). In a recent study, when a high dose of cisplatin was used, some mice died because of the general toxicity of cisplatin; however, when mice were pretreated by UTL-5g, it was found that UTL-5g not only increased the survival rate but also delayed the time to death [73]. These studies further support that UTL-5g is a potential broad-spectrum chemoprotector.

The other two approved platinum drugs, carboplatin and oxaliplatin, are also associated with side effects similar to those of cisplatin although somewhat different. Nevertheless, the important side effect, myelosuppression, remains the same for all these platinum drugs. Some of the UTL-5g studies indicate that it also possess similar protective effects against the side effects-induced by carboplatin or oxaliplatin.

Normal Blood Cell Counts in human	
Red Blood Cells	4.7 - 6.1 x 10^6 per μL
White Blood Cells	5,000 - 10,000 per μL
Platelets	150,000 - 400,000 per μL

Part III

Chapter 3
Drugs and Potential Drugs as Radiosensitizers, Radioprotectors, or Radiomitigators

As mentioned in previous sections, there are three types of adjuvant drugs that could be used in conjunction with radiotherapy. In this section, we will describe some drugs and potential drugs in each of these three categories: radiosensitizers, radioprotectors, and radiomitigators.

1. RADIOSENSITIZERS

Currently, there are a few agents under investigation as radiosensitizers. Some examples are shown below:

Tirapazamine (SR-4233)

Tirapazamine (3-amino-1,2,4-benzotriazine-1,4-dioxide, Fig. 3.6) was originally identified as an anticancer agent by SRI International and Stanford University researchers investigating oxygen deficiency (hypoxia) in solid tumors. It is an experimental anticancer drug that is supposed to be activated under low levels of oxygen. In other words, it is activated to its toxic form preferentially in the hypoxic areas of solid tumors and induces DNA damage and cell death. A recent report on tirapazamine indicated that selective targeting of hypoxic cells in primary tumors is a viable approach in controlling metastatic disease [74]. In addition, tirapazamine also works as a radiosensitizer and makes the hypoxic cells more sensitive to radiation therapy. A number of clinical studies have been conducted in patients with head and neck cancer, gynecological cancer, and other solid tumor types [75, 76].

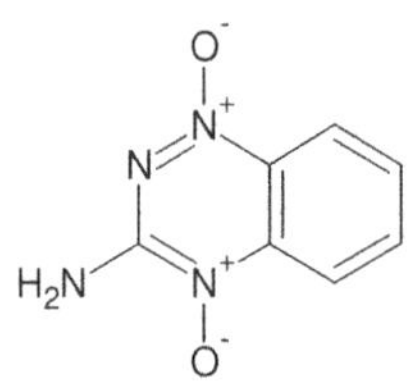

Fig. 3.6 Tirapazamine

DCQ (2-benzoyl-3-phenyl-6,7-dichloroquinoxaline 1,4-dioxide)

DCQ (2-benzoyl-3-phenyl-6,7-dichloroquinoxaline 1,4-dioxide, Fig. 3.7) is a synthetic quinoxaline 1,4-dioxide. It was reported to enhance the cytotoxic effect of ionizing radiation. Itani et al. showed that upon treatment of four cancer cell lines, DCQ is effective as a hypoxic cell radiosensitizer on one cancer cell line (DLD-1 human colon cancer) [77]. The mechanism of action was reported as mediated by increasing DNA damage and/or decreasing DNA repair and that ROS are at least partially responsible [78].

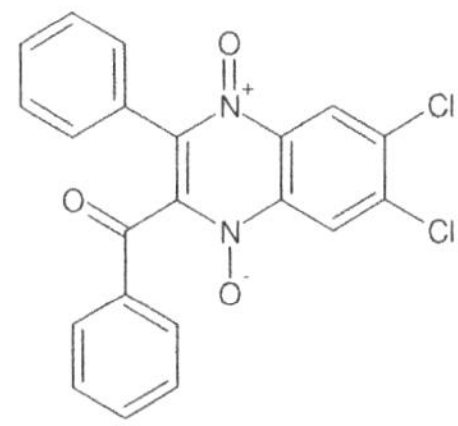

Fig. 3.7 DCQ

Misonidazole or 1-(alpha-methoxymethylethanol)-2-Nitroimidazole

Misonidazole (Fig. 3.8) is a nitroimidazole with radiosensitizing and antineoplastic properties. Misonidazole is used as a hypoxic cell sensitizer in radiation therapy to cause normally resistant hypoxic tumor cells to become sensitive to radiotherapy. Its high electron affinity helps induce the formation of free radicals, thereby sensitizing hypoxic cells to cell killing effect of ionizing radiation.

A number of phase III clinical studies have been conducted utilizing misonidazole. However, not all studies are positive. For example, a report in 1983 showed that 438 patients were treated between 1979 and 1981 in a double-blind randomized study to evaluate the effect of misonidazole in conjunction with radiotherapy on grades 3 and 4 astrocytoma. The results of this study did not show any significant positive effect of misonidazole [79].

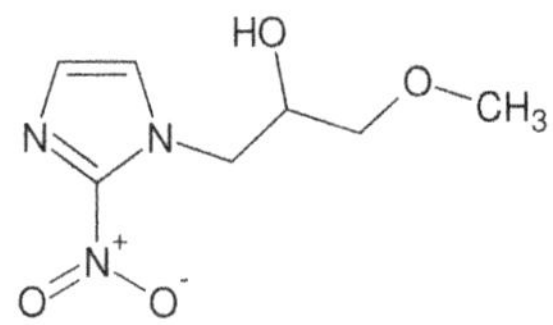

Fig. 3.8 Structure of misonidazole

Co(BCA) or nitro-bis(2,4-pentanedionato)(bis(2-chloroethyl)amine)cobalt(III)

Co(BCA) is a Co(III) complex (Figure 3.9) being investigated as a potential radiosensitizer and antitumor agent [80]. Studies by Teicher et al. indicate that Co(BCA) has significant antineoplastic effects *in vitro* and *in vivo* and Co(BCA) interacts positively with both radiation and mild hyperthermia [80]. The broad therapeutic dose range of Co(BCA) further suggests its potential clinical utility.

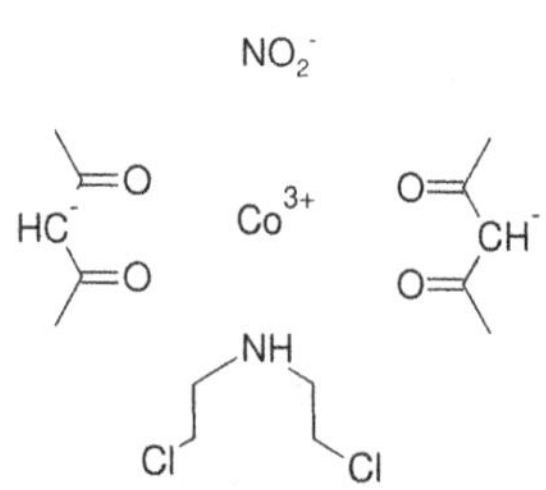

Fig. 3.9 Co(BCA)

Meloxicam (Sold as Mobic in the US)

Overexpression of cyclooxygenase-2 protein (COX-2) is often associated with some cancers, including glioma. Therefore, COX-2 is a potential target for cancer prevention and treatment.

Meloxicam (Fig. 3.10) is a non-steroidal anti-inflammatory drug (NSAID) that works as an inhibitor of COX-2 [81]. Studies on the effects of meloxicam alone and in combination with radiation on human glioma cells indicated that meloxicam exerted COX-2 independent growth inhibition and radiosensitization of human glioma cells [82].

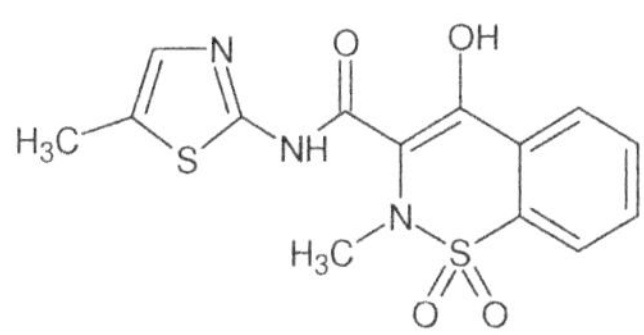

Fig. 3.10 Structure of meloxicam

2. RADIOPROTECTORS

In general, radioprotectors (radioprotective agents) can be categorized into the following classes:

A. Thio Compounds

Amifostine

As a radioprotector, Ethyol® is first reconstituted by saline and then administered through infusion intravenously. Infusion delivers amifostine in a lower concentration over a longer period of time (it can take up to 15 minutes) so that the potential side effects of the drug can be reduced. However, infusion is inconvenient as compared to a normal injection or oral administration.

Once administered, amifostine is metabolized by phosphatase and becomes a free sulfhydryl (thiol), which can scavenge the free radicals induced by radiation and ionizing radiation to achieve the protection effect. The selective protection of normal tissues is not completely understood, but it is generally believed to be due to higher alkaline phosphatase activity, higher pH, and vascular permeation of normal tissues. For example, alkaline phosphatase is present in all organs, especially in liver, kidney, and bone, thus, facilitating the broad-spectrum protection by amifostine.

Amifostine has been used to reduce the side effects induced by DNA-damaging radiotherapy and DNA-binding chemotherapeutic agents including platinum drugs (e.g. cisplatin) for a number of years. Unfortunately, amifostine has several shortcomings.

Treatment with amifostine resulted in nausea, vomiting, diarrhea, and transit hypotension [72]. While the use of amifostine is becoming more widespread, an increased number of cutaneous reactions have also been reported [83]. One report suggested that in rats, amifostine is behaviorally toxic at doses relevant to radioprotection [84]. In addition, amifostine combined with radiation disrupts performance more severely than either radiation or amifostine alone [84]. Therefore, amifostine does not appear to be an ideal radioprotector although amifostine was reported to provide significant protection against cyclophosphamide-induced hematologic toxicity [85].

Sagowski et al. reported that amifostine has a significant cytoprotective effect on the late toxicity of irradiated salivary glands in their study on rats. However, the acute toxicity of the glands during radiotherapy was not reduced in their study [86]. In addition, while protecting normal cells, amifostine has also been reported to actually protect tumor cells in some animal studies [66].

<u>WR-3689</u>

Another aminothiol compound, WR-3689 or S-2-(3-methylaminopropylamino)ethylphosphorothioic acid, was reported to be effective in radioprotection. As shown in Fig. 3.11, WR-3689 and amifostine are strikingly similar, with only one additional methyl group at the amino end, thus its radioprotective effect is not a total surprise. WR-3689 showed protection against gamma-ray-induced intestinal damage [87].

When used in combination with vitamin E before irradiation, the radiation dose reduction factor (DRF) and the survival in mice were significantly increased [60].

Cystamine

Cystaminum (cystamine)(Fig. 3.11), was approved in Russia as a radioprotector. Other aminothiols have been shown to be effective in radioprotection. It is injected 10 - 30 min before radiotherapy and the protective effect lasts about 5 hours.

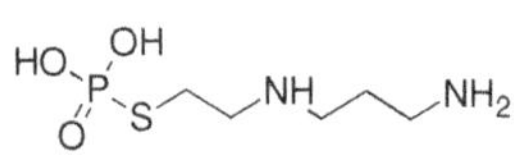

Amifostine (WR-2721)

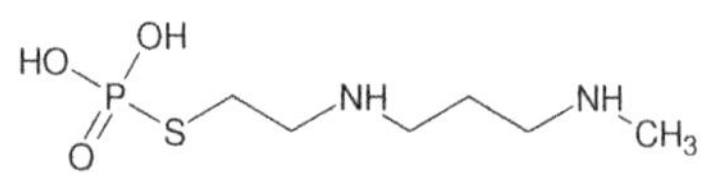

WR-3689

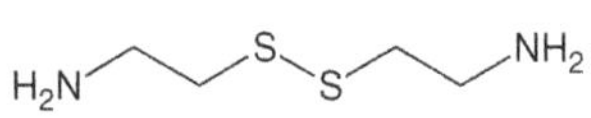

Cystaminum (Cystamine)

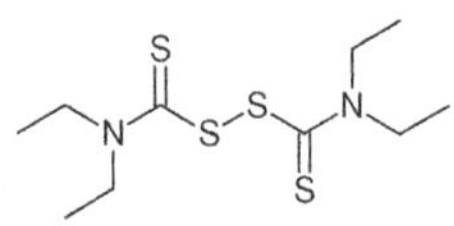

Disulfiram (tetraethylthiuramdisulphide)

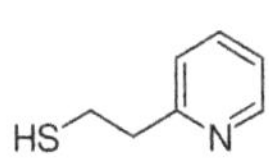

2-pyridinemethanethiol

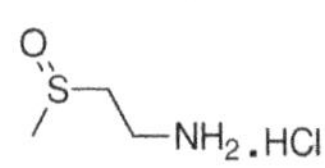

2-(methylsulfinyl)ethylamine

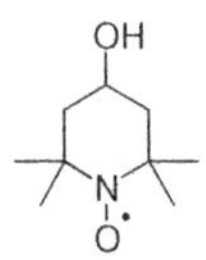

tempol (4-hydroxy-2,2,6,6,tetramethylpiperidine-1-oxyl)

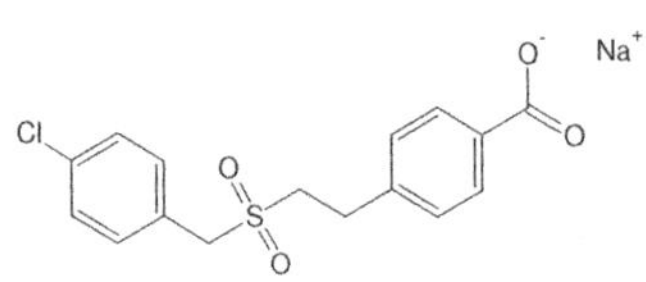

Recilisib

Fig. 3.11 Amifostine and other potential radioprotective agents

Other thio compounds

In addition, a number of pyridine and pyrimidine containing sulfur side-group have been investigated as potential radioprotectors. For example, 2-pyridinemethanethiol (Fig. 3.11), has been studied as a potential radioprotector, but with limited effect[88].

Sulfoxides have also been investigated. For example, 2-(methylsulfinyl)ethylamine hydrochloride (Figure 3.11) was studied, but with only limited radioprotective effect [89].

Recilisib (Fig. 3.11) is a newer thiol compound (Ex-Rad® by Onconova Therapeutics) which is a small-molecule radioprotective agent. It is being developed to treat radiation-induced cytopenia, which is a reduction in the number of blood cells.

Recilisib works by up-regulation of phosphatidylinositide 3-kinases (PI 3-kinases)/AKT (Protein Kinnase B) pathways to reduce DNA damage in cells exposed to radiation [90].

In some cancers, this pathway is overactive, thus reducing apoptosis and allowing proliferation. By up-regulating the PI 3-kinase/AKT pathway, recilisib should reduce apoptosis induced by radiation.

B. Nitroxides

Other than thiol compounds, nitroxides may be the second most promising group of compounds as potential radioprotectors and some are in clinical development [91]. Nitroxide free radicals and their reduction product, hydroxyl amine, are recycling antioxidants, but only the nitroxide free radicals are radioprotective [60]. Currently, the most advanced nitroxide under development is tempol (4-hydroxy-2,2,6,6-tetramethylpiperidine-1-oxyl) (Figure 3.12) [92, 93].

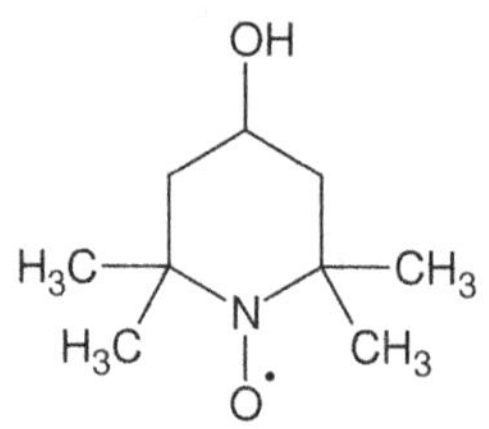

Fig. 3.12 Structure of tempo (4-hydroxy-TEMPO)

Preclinical studies showed that topical application of tempol was effective in preventing radiation-induced alopecia in guinea pigs [94] and in human phase I clinical study [95]. Tempol was also shown to protect salivary gland against radiation-induced damage and does not alter tumor growth [96].

C. Polysaccharides

Another type of compounds, low toxic microbial polysaccharides, has been shown to exhibit radioprotective and immunomodulating activity. For example, the *i.v.* injection of several individual polysaccharides (60 mg/kg) significantly decreased the mortality of mice exposed to a single dose of x rays [97]. Glucan, a beta-1,3-polysaccharide immunomodulator, was shown to enhance the survival of mice when administered before radiation exposure. Pre-irradiation treatment of glucan in combination with WR-2721 (amifostine) or WR-3689 synergistically increases both hematopoietic reconstitution and survival [98].

D. Morpholino Oligonucleotides (Antisense)

A more recent discovery by Isenberg et al. showed that a morpholino antisense oligonucleotide (ASO) targeting CD47 confers the radioresistance on human endothelial cells *in vitro* and protects soft tissue, bone marrow, and tumor-associated leukocytes in irradiated mice. It was shown that thrombospondin-1 (THBS1) signaling through its cell surface receptor CD47 limits recovery from several types of stress, and mice lacking either gene are profoundly resistant to radiation injury[99].

Lawrence and coworkers indicated found that radiation-induced liver apoptosis and hepatocyte micronucleus formation were reduced by pretreatment with ASO against TNF receptor 1 (TNFR1) indicating that ionizing radiation activates apoptotic signaling through TNFR1 in the liver [100].

E. Other potential radioprotective agents

A number of other potential radioprotectors have been reported and some examples are shown below:

(a) CBLB502 is a polypeptide that binds and activates Toll-like receptor 5 (TLR 5), which is expressed on enterocytes [101] and intestinal endothelial cells [60]. Activation of TLR 5 results in activation of nuclear factor κB (NFκB), which plays an important role in the response of the intestine to irradiation [60]. Studies showed that CBLB502 protected mice and monkeys from lethal TBI [102].

(b) Melatonin was shown to be radioprotective in several animal studies [103-105]. Melatonin has also been described as a radiosensitizer in animal models[106].

(c) Eumelanin was reported to be radioprotective and may have potential benefit for the radiotherapy of malignant melanoma[107].

(d) Vitamins A, C, and E have been studied as potential radioprotectors. For example, vitamin E was found to be radioprotective especially when used in combination with other radioprotective agent such as WR-3689[108]. However, vitamins are generally not as effective as other radioprotectors.

F. Our Unique Radioprotectors

It has been reported that TNF-α was induced by radiation during the acute phase of pneumonitis and the predominant localization of TNF-α in areas of inflammatory cell infiltrates suggests involvement of TNF-α in the pathogenesis of radiation-induced lung injury [109]. Another report suggested that the radioprotective effects of IL-11 may be mediated via the modulation of TNF-α production [110]. Yet another study showed that caffeic acid phenethyl ester (CAPE) treatment decreased the expression of inflammatory cytokines including TNF-α after irradiation and decreased radiation-induced interstitial pneumonitis and TGF-β expression[111]. Therefore, it is reasonable to hypothesize that certain TNF-α modulators may be lung radioprotective.

We have developed a series of small-molecule TNF-α modulators, UTL-5 series, as potential radioprotectors. Originally, UTL-5 series of compounds were developed as potential immunomodulators. Based on the work conducted to date, three of them (UTL-5b, -5d, and 5g shown in Fig. 3.13) are TNF-α modulators and have shown promising activities in anti-inflammation, anti-arthritis, liver radioprotection, and chemoprotection [112-114]. Although these three compounds behave similarly, based on the data available, we have selected UTL-5g as the lead compound for lung cancer radioprotection and the preliminary animal results indicated that UTL-5g may be lung radioprotective [115]. Our recent studies showed that it increased the survival rate for mice underwent TBI and increased WBC count reduced by the radiation.

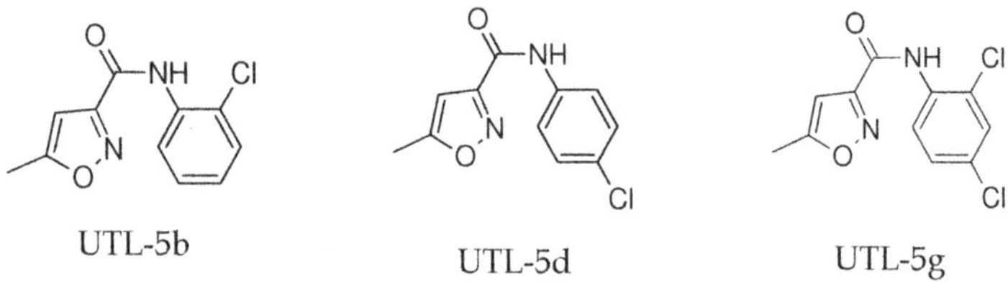

Fig. 3.13 Structures of UTL-5b, -5d, and -5g

In an animal study, mice were pre-treated with UTL-5g prior to a TBI at 5 Gy followed with 4 additional daily UTL-5g treatments. Ten days later, spleen colonies were examined. As shown in Fig. 3.14, spleen from UTL-5g-treated mice showed more and larger endogenous spleen colonies compared to those without treatment, indicating that UTL-5g can protect and promote the recovery of hematopoietic stem cells in irradiated mice.

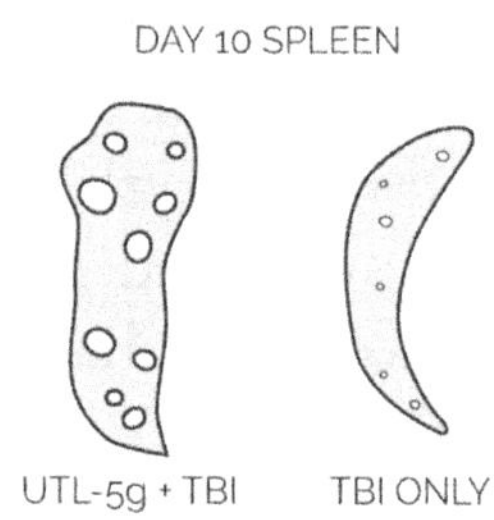

Fig. 3.14 Spleens from mice under total body irradiation with UTL-5g (left) and without UTL-5g pretreatment

Recently, we studied a new series of small molecules, UTS-series (patent pending), and found that they significantly increased the survival rates of NIH Swiss mice under 7.5 Gy TBI. Therefore, UTS compounds may be both radioprotective and chemoprotective. Further research work is being conducted.

3. POTENTIAL RADIOMITIGATORS

Discovery and development of radiomitigators are becoming increasingly important. This is because radiation can induce immune depression and development of related pathologies. The other reason is national defense.

Rx100

Rx100 (Fig. 3.15) is a sulfur containing small-molecule under development as a potential radioprotector. It is a lysophosphatidic acid (LPA) receptor modulator. LPA activates several G-protein-coupled receptors, including LPA1 through LAP6. Because LAPs ares able to stimulate cell proliferation, dysregulation of LPA receptors or autotaxin (a lysophospholipase that regulates LPA production) may be linked to cancer.

Rx100 has been shown to be a potential agent for both radioprotection and radiomitigation. It is now under investigation for gastro-intestinal acute radiation syndrome (GI-ARS).

In the US patent #8,686,177, 4/1/2014, a series of molecules, including Rx100, were claimed. The key molecular component of Rx100 and relative compounds, O-substituted dihydrogen phosphorothioate, is as follows, wherein R represents a variety of groups.

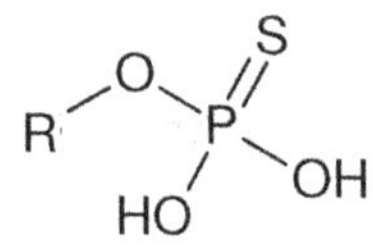

Fig. 3.15 Basic structure of Rx100 and related compounds

KZ-41

KZ-41 (Fig. 3.16) is patented (US patent 8,115,031, 2012), in that a general structure of this type of quinic acid derivatives is disclosed as follows:

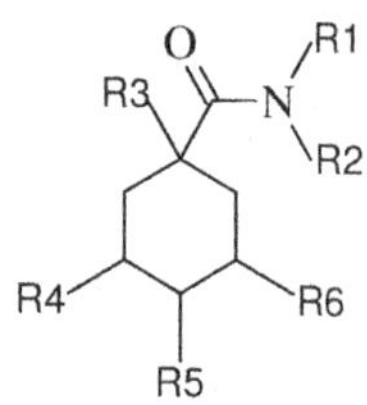

Fig. 3.16 Structure of KZ-41

wherein R1 and R2 are each independently hydrogen, straight or branched alkyl, cycloalkyl, aryl, benzyl, arylalkyl, heterocyclic amine, aminoalkyl, ketone, or

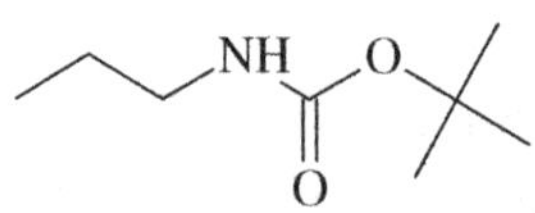

and R3, R4, R5, and R6 are each independently hydrogen or hydroxyl.

KZ-41 is a quinic acid derivative under development as an orally available potential mitigator for radiation injury [116]. It works through inhibition of p38 MAPK-p53-ICAM-1 signaling [117].

KZ-41 shows radiomitigating activity in lethally irradiated mice. Imaging results suggest the radiomitigating activity of KZ-41 is through mechanisms involving promotion of initial clot formation and vascular flow restoration [118]. Another publication showed that KZ-41 prevents retinal endothelial cell apoptosis through p38 MAPK-dependent pathways [108].

ARA290

ARA290 is a synthetic peptide under investigation as a potential radiomitigator for both hematopoietic and gastrointestinal syndromes of the acute radiation syndrome. In animal studies, it was reported to stop the spread of inflammation and cellular death from sites of radiation injury into healthy surrounding tissues [119].

TGF-β inhibitor

TGF-β plays an important role in the radiation-induced fibrosis. Therefore, inhibition of TGF-β is of interest in the research of radiation mitigation. A rat study showed that TGF-β1 receptor inhibitor, SM16, at a daily dose of 150 mg/kg for 3 weeks reduced functional lung damage, morphologic changes, and inflammatory response at 26 weeks after radiotherapy [92].

A small-molecule plant derived flavonoid, halofuginone (Fig. 3.17), is a TGF-β inhibitor; recent studies showed halofuginone ameliorated radiation-induced soft tissue fibrosis [120]. In a mouse study, halofuginone treatment by interperitoneal (*i.p.*) injection significantly lessened radiation-induced fibrosis through the modulation of TGF-β signal pathway [121]. However, halofuginone does not reduce fibrosis in bleomycin-induced lung injury [122].

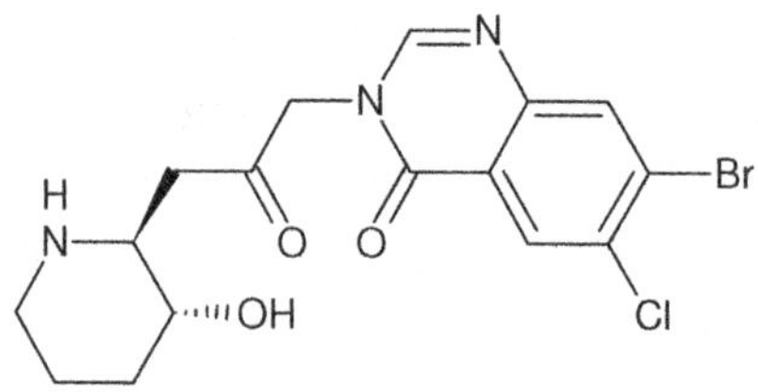

Fig. 3.17 Structure of halofuginone

Part III

Chapter 4
Use of Natural Products

In the earlier chapter, we mentioned that bone marrow and intestinal damage are among the important side effects of radiotherapy. If the whole body is exposed to radiation, as might be the case in a nuclear accident or terrorist incident, it could result in death or cancer. In addition to approved drugs or agents under investigation, a number of natural products have been reported to be useful in dealing with the side effects resulted from chemotherapy and/or radiotherapy; some of these natural products have been commercialized and sold as dietary supplements.

Although many natural products have been reported anecdotally to be beneficial to patients in many different therapeutic areas, only a few have made it through the stringent process of drug development and approved by the FDA as official drugs in different therapeutic areas. Due to the strong interest in nature products in recent years, there has been more and more research done and published for natural products with medicinal effects.

To date, none of the natural products have been approved as chemoprotective or radioprotective drugs by the FDA even though some natural products have been reported to be effective in chemoprotection or radioprotection.

One word of caution, dietary supplements may be associated with a wide range of toxicity. For example, a popular dietary supplement, milk thistle, is supposed to protects liver, but it could also cause some side effects including nausea and bloating. The potential side effects of each dietary supplement may vary and most of them are still not thoroughly investigated.

The following are some representative examples of natural products, which are potentially useful in chemoprotection and/or radioprotection. These natural products are divided into two categories: (A) naturally occurring compounds, and (B) extracts (comprising more than one compound) of natural products.

1. Naturally Occurring Compounds

Vitamins

Some vitamins are reported to have certain radioprotective effect. For example, vitamin E is a family of important fat-soluble antioxidants, including α-, β-, γ-, and δ-tocopherols and four corresponding tocotrienols. High levels of vitamin E can be found in a variety of food including egg, wheat germ, nuts, milk, and seeds.

Konopacka et al. reported that pretreatment of vitamin E (100-200 mg/kg/day) and β-carotene (3-12 mg/kg/day) were effective in protecting mice against damage induced by gamma rays. The most effective protection was noted when a mixture of vitamin E, β-carotene, and vitamin C was used as a pretreatment [123].

Wadleigh et al. conducted a randomized, double-blind, placebo-controlled study to evaluate the effect of topical vitamin E in the treatment of oral mucositis in patients receiving chemotherapy for various types of malignancy. Six of the patients receiving vitamin E had complete resolution of their oral lesions. Their conclusion is that vitamin E may be an effective therapy in patients with chemotherapy-induced mucositis [124].

In another investigation Kottschade et al. conducted a phase III, double-blind, placebo controlled study, to evaluate the use of vitamin E for the prevention of chemotherapy induced peripheral neuropathy [125], which is the damage to peripheral nervous system. However, they concluded that Vitamin E did not appear to reduce the incidence of sensory neuropathy in the studied group of patients receiving neurotoxic chemotherapy except that vitamin E appeared to be well tolerated.

Kumar et al. conducted an *in vitro* study on three human fibroblast cell lines and three human cancer cell lines, cervical cancer (HeLa) and ovarian carcinoma cells (OVG1 and SKOV3). They pre-treated these cells with α-tocopherol succinate (37.6 μM) 20 hours before 1 Gy γ-radiation. Their results showed that α-tocopherol succinate induced chromosomal damage in both human cervical cancer cells and ovarian cancer cells, but not in human normal fibroblasts in culture. In addition, α-tocopherol succinate protected normal fibroblast cells from radiation-induced chromosomal damage [126].

Although some vitamins do show beneficial effects in reducing the side effects of chemotherapy and/or radiotherapy, one needs to be careful in taking vitamins at higher doses because it can produce some toxicities. For example, vitamin E at doses 2,000 IU or more after prolonged consumption may induce clotting defects. Beta-carotene at doses of 60 mg or more can induce bronzing of skin although it is reversible after discontinuing the intake [127].

Hesperidin

Hesperidin (Fig. 3.18) is a flavonoid found abundantly in the peel of citrus fruits, such as orange, lime, and lemon. Flavonoids (sometimes also referred to as bioflavonoids) are widely distributed in different plants for many functions. They are most commonly known for their antioxidant activity.

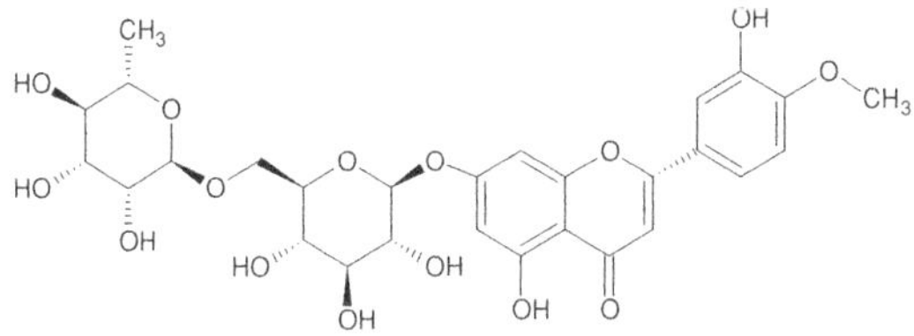

Fig. 3.18 Molecular structure of hesperidin

Hesperidin is an antioxidant and was reported to be able to penetrate the blood brain barrier in an *in vitro* model [60]. Based on a paper in British Journal of Radiology in 2006, hesperidin was demonstrated to have powerful protective effects against the radiation-induced DNA damage and the decline in cell proliferation in mouse bone marrow [60].

A recent scientific article reported the preventive effect of hesperidin in mouse bone marrow cells against genotoxicty induced by cyclophosphamide [128]. The article concluded that hesperidin reduced the oxidative stress and genotoxicity induced by cyclophosphamide in mouse bone marrow cells.

Commercial products (in capsule form) made from hesperidin are available from several companies as dietary supplements.

Superoxide Dismutase (SOD)

SOD is a class of intracellular enzymes first discovered by McCord and Fridovich [129]. SOD is considered as an antioxidant because it scavenges the superoxide, hydroxyl, and other oxygenated free radicals [130]. Several attempts have been made to use

SOD as a radioprotector in patients receiving radiotherapy. However, the results were controversial [131-133].

N-acetylcysteine (NAC)

NAC (Fig. 3.19) is a nutritional supplement that has been available for a number of years. It is also sold as a prescription drug. For example, inhaled NAC is used as a mucolytic agent, which is an agent that dissolves thick mucus to help relieve respiratory difficulties. It was reported that NAC may be used for the prevention of radiocontrast-induced kidney injury. Marenzi et al. reported that NAC may prevent contrast-medium-induced nephropathy with a dose-dependent effect in patients treated with primary angioplasty and may improve hospital outcome [134].

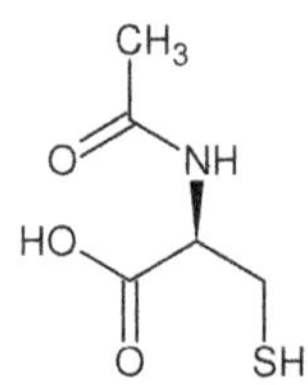

Fig. 3.19 Molecular structure of N-acetyl cysteine

Beta Glucan

Beta glucan is a polysaccharide of D-glucose monomers linked by glycosidic bonds. It was reported that beta glucan, a beta-1,3-polysaccharide immunomodulator, enhances survival of mice when administered before radiation exposure. Therapeutic glucan treatment administered in combination with pre-irradiation treatment of WR-2721 or WR-3689 synergistically increases both hematopoietic reconstitution and survival [98].

Quercetin

Quercetin (Fig. 3.20) is a naturally occurring flavonoid. Studies suggest that quercetin, when given as pre-treatment, protected against radiation-induced primary leukocyte DNA damage in mice and damage in human peripheral blood lymphocytes [60, 135].

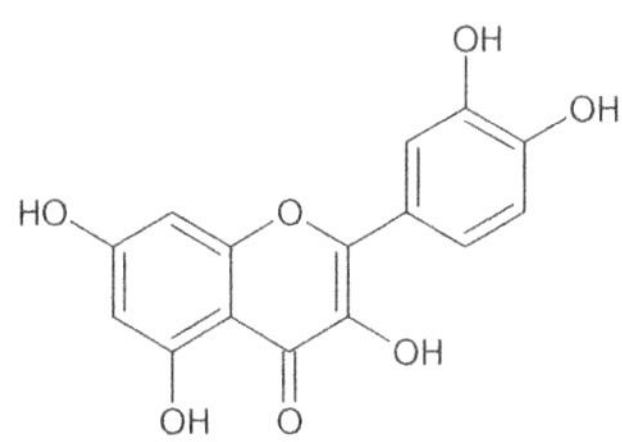

Fig. 3.20 Structure of quercetin

2. Extracts of Natural Products

The following are examples of extracts from natural products. Please note, extracts of natural products comprise many compounds and they are different from naturally occurring individual compounds. In addition, because they are generally not regulated as drugs are, consumers must be cautious about the potential quality issues.

Cordyceps Sinensis ("Dong tsong shia tsao" in mandarin)

Cordyceps sinensis (CS) is a fungal parasite of moth larvae grown in Tibet (Fig. 3.21). "Dong tsong shia tsao" in Chinese means "worm in winter and grass in summer." Because of the way it looks, it is also called caterpillar fungus.

CS is a novel traditional Chinese medicine supposedly effective in modulating immune system and improving general health. Liu and co-workers reported that an orally administered extract of CS protects mice from bone marrow and intestinal injuries after TBI [136]. Their studies showed that CS increased the median time to death from 13 to 20 days after 8 Gy TBI and from 9 to 18 days after 10 Gy TBI. At lower TBI doses of 5.5 and 6.5 Gy, CS protected against bone marrow death, which was further confirmed by the more rapidly recovery of WBC counts.

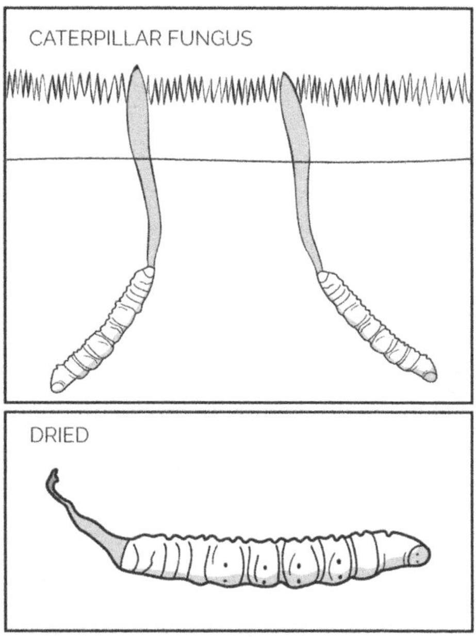

Fig. 3.21 Fresh (top) and dried (bottom) *Cordyceps sinensis*

CS has been widely considered a noble herb in China and is very expensive. Although the commercial products made from CS are available, it has been reported that many times the products do not actually contain much CS. Therefore, consumers beware!

Ocimum Sanctum

Ocimum is a genus of aromatic annual and perennial herbs, often used for religious and medicinal purposes; these plants have been widely used in South Asia as an herbal tea. A scientific paper published in *British Journal of Radiology* indicated that two flavonoids, orientin (Ot) and vicenin(Vc), obtained from the leaves of Ocimum sanctum showed radioprotective effectiveness in mice and concluded that both flavonoids may be promising for human radiation protection [137].

Radix Glycyrrhiza (also referred to as "Gang Tsao" in Chinese)

Radix Glycyrrhiza is a popular traditional Chinese herb. It has a special sweet taste as implied by its Chinese name, Gang Tsao, which literally means sweet grass.

Active ingredients from Radix Glycyrrhiza include glycyrrhizin, glycyrrhetinic acid, liquiritin and liquiritigenin. Traditionally, it has been used as a detoxifying agent in China. Glycyrrhizin has been shown to possess many physiological functions including anti-inflammatory activity, detoxification and inhibition of carcinogenic promoters. Rahman et al. studied the chemopreventive effect of glycyrrhizin on TPA (12-*O*-tetradecanoyl phorbol-13-acetate), which is a phorbal ester known for its tumor promotion activity. The induction of skin inflammation by TPA is believed to be governed by cyclooxygenase (COX), lipoxygenase, and ornithine decarboxylase (ODC), all are important markers for skin tumor promotion. They reported that pretreatment with glycyrrhizin is protective against TPA-induced oxidative stress and tumor promotion in Swiss albino mice [138].

<u>Milk Thistle</u>

Milk thistle (Fig. 3.22) belongs to the daisy family. If you have a lawn, you may have seen it with sharp prickles on the leaf margins. Normally, milk thistle extract is made from the seeds, which contain approximately 4 - 6% silymarin (Fig. 3.23). Silymarin is a mixture of flavonolignana, consisting of silibinin isosibilinin, silicristin, and silidian [139]. Silibinin has been reported to have protective effect on liver against toxins [140].

Fig. 3.22 Milk thistle

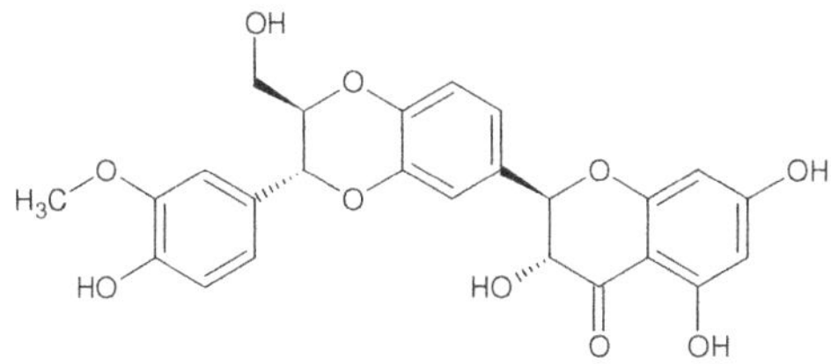

Fig. 3.23 Structure of silibinin

Milk thistle has been used as a medicinal herb for many years to treat chronic liver disease [141, 142]. It has been investigated for liver protection from exposure to certain chemicals [143].

Ganoderma

Ganoderma (Fig. 3.24) is a genus of polypore mushrooms which are wood-decaying fungi. It is one of the most popular dietary supplements in China and believed to be useful for many diseases thus it is also call lingzhi. In Chinese, lingzi means wonderful and magnificent. It is not a plant, but a fungus which extracts nutrients from its hosting plant.

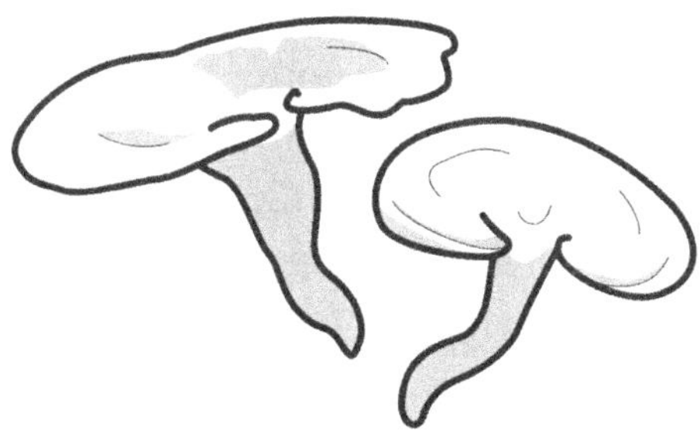

Fig. 3.24 Ganoderma

Ganoderma species have been intensely investigated for a variety of potential therapeutic benefits including antioxidant and anticancer effects, acting supposedly by strengthening the immune system and having liver-protecting effects [144-146]. Commercial products as dietary supplements are also available.

Peppermint (Mentha piperita)

It has been reported that, in animal studies, peppermint leaf-extract has a significant radioprotective effect, which may be attributable to the antioxidant and radical scavenging activities of the flavonoids and flavonols in the extract [147, 148].

QR-336 (an herbal extract)

QR-336 is a standardized concentrated nutrient/phytochemical based compound that has been shown in mouse studies to reduce the side effects of ionizing radiation when administered *i.p.*, suncutaneous (*s.c.*) or orally prior to radiation exposure indicating that QR-336 may be a potential radioprotective agent. Pretreatment of QR-336 protected mice in a lethal-dose radiation study.

Antrodia Cinnamomea

Antrodia cinnamomea is a unique fungus grown mostly in high moisture mountain areas in Taiwan. They are usually red or yellow in color. White Antrodia cinnamomea is rare, thus, very expensive primarily due to its effects in anticancer and liver detoxification.

However, not all studies indicate the beneficial pharmacological effects; some even showed kidney toxicity. Therefore, it is somewhat controversial as to its health benefits.

The active ingredients of Antrodia cinnamomea include polysaccharides, triterpenoids, and superoxide dismutase. Currently, there are commercial products available based on the extract of Antrodia cinnamomea.

References

1. McGovern, P.C., et al., *Risk factors for postcraniotomy surgical site infection after 1,3-bis (2-chloroethyl)-1-nitrosourea (Gliadel) wafer placement.* Clin Infect Dis, 2003. **36**(6): p. 759-65.
2. Rosenberg, B., L. Vancamp, and T. Krigas, *Inhibition of Cell Division in Escherichia Coli by Electrolysis Products from a Platinum Electrode.* Nature, 1965. **205**: p. 698-9.
3. Shaw, J., *Cisplatin analogs for cancer treatment* U. PTO, Editor 1999: U.S.
4. Shaw, J., *Dinuclear platinum complexes as cisplatin analogs for cancer treatment* U. PTO, Editor 2000: U.S.
5. Messori, L., et al., *Structural features of a new dinuclear platinum(II) complex with significant antiproliferative activity.* Inorg Chem, 2003. **42**(20): p. 6166-8.
6. Peters, G.J., et al., *Basis for effective combination cancer chemotherapy with antimetabolites.* Pharmacol Ther, 2000. **87**(2-3): p. 227-53.
7. Christensen, S. *Why Can't I Take Folic Acid the Same Day As Methotrexate?* 2014; Available from: http://www.livestrong.com/article/420615-why-cant-i-take-folic-acid-the-same-day-as-methotrexate/.
8. Noble, R.L., C.T. Beer, and J.H. Cutts, *Role of chance observations in chemotherapy: Vinca rosea.* Ann N Y Acad Sci, 1958. **76**(3): p. 882-94.
9. Grunberg, S.M., et al., *Treatment of non-small-cell lung cancer with vinblastine and very high-dose cisplatin. A Southwest Oncology Group study.* Cancer Chemother Pharmacol, 1991. **28**(3): p. 211-3.
10. Bellmunt, J., et al., *Carboplatin, methotrexate, and vinblastine in patients with bladder cancer who were ineligible for cisplatin-based chemotherapy.* Cancer, 1992. **70**(7): p. 1974-9.
11. Keating, G.M., *Afatinib: a review of its use in the treatment of advanced non-small cell lung cancer.* Drugs, 2014. **74**(2): p. 207-21.
12. Miller, V.A., et al., *Afatinib versus placebo for patients with advanced, metastatic non-small-cell lung cancer after failure of erlotinib, gefitinib, or both, and one or two lines of chemotherapy (LUX-Lung 1): a phase 2b/3 randomised trial.* Lancet Oncol, 2012. **13**(5): p. 528-38.
13. Rich, J.N., B.K. Rasheed, and H. Yan, *EGFR mutations and sensitivity to gefitinib.* N Engl J Med, 2004. **351**(12): p. 1260-1; author reply 1260-1.
14. Lin, N.U., et al., *A phase II study of afatinib (BIBW 2992), an irreversible ErbB family blocker, in patients with HER2-positive metastatic breast cancer progressing after trastuzumab.* Breast Cancer Res Treat, 2012. **133**(3): p. 1057-65.
15. Lissoni, P., et al., *Modulation of cancer endocrine therapy by melatonin: a phase II study of tamoxifen plus melatonin in metastatic breast cancer patients progressing under tamoxifen alone.* Br J Cancer, 1995. **71**(4): p. 854-6.
16. Desta, Z., et al., *Comprehensive evaluation of tamoxifen sequential biotransformation by the human cytochrome P450 system in vitro: prominent roles for CYP3A and CYP2D6.* J Pharmacol Exp Ther, 2004. **310**(3): p. 1062-75.
17. Jordan, V.C., *Fourteenth Gaddum Memorial Lecture. A current view of tamoxifen for the treatment and prevention of breast cancer.* Br J Pharmacol, 1993. **110**(2): p. 507-17.

18. FDA. *Tamoxifen*. Congressional Testimony 1998 1998 [cited 2013 December 28]; Available from: http://www.fda.gov/newsevents/testimony/ucm115118.htm.
19. Miller, R.P., et al., *Mechanisms of Cisplatin nephrotoxicity*. Toxins (Basel), 2010. **2**(11): p. 2490-518.
20. Madias, N.E. and J.T. Harrington, *Platinum nephrotoxicity*. Am J Med, 1978. **65**(2): p. 307-14.
21. Koski, L., et al., *Modulation of cortical activity during different imitative behaviors*. J Neurophysiol, 2003. **89**(1): p. 460-71.
22. Alexander, J., et al., *Serial assessment of doxorubicin cardiotoxicity with quantitative radionuclide angiocardiography*. N Engl J Med, 1979. **300**(6): p. 278-83.
23. Buzdar, A.U., et al., *Early and delayed clinical cardiotoxicity of doxorubicin*. Cancer, 1985. **55**(12): p. 2761-5.
24. Naidu, M.U., et al., *Chemotherapy-induced and/or radiation therapy-induced oral mucositis--complicating the treatment of cancer*. Neoplasia, 2004. **6**(5): p. 423-31.
25. Wolf, S., et al., *Chemotherapy-induced peripheral neuropathy: prevention and treatment strategies*. Eur J Cancer, 2008. **44**(11): p. 1507-15.
26. Chaudhry, V., et al., *Peripheral neuropathy from taxol and cisplatin combination chemotherapy: clinical and electrophysiological studies*. Ann Neurol, 1994. **35**(3): p. 304-11.
27. Podratz, J.L., et al., *Cisplatin induced mitochondrial DNA damage in dorsal root ganglion neurons*. Neurobiol Dis, 2011. **41**(3): p. 661-8.
28. Areti, A., et al., *Oxidative stress and nerve damage: role in chemotherapy induced peripheral neuropathy*. Redox Biol, 2014. **2**: p. 289-95.
29. Hueso, L., et al., *[Chemotherapy-induced acral erythema: a clinical and histopathologic study of 44 cases]*. Actas Dermosifiliogr, 2008. **99**(4): p. 281-90.
30. Rybak, L.P., et al., *Cisplatin ototoxicity and protection: clinical and experimental studies*. Tohoku J Exp Med, 2009. **219**(3): p. 177-86.
31. Pinkel, D., et al., *Radiotherapy in leukemia and lymphoma of children*. Cancer, 1977. **39**(2 Suppl): p. 817-24.
32. Thomas, E.D., et al., *One hundred patients with acute leukemia treated by chemotherapy, total body irradiation, and allogeneic marrow transplantation*. Blood, 1977. **49**(4): p. 511-33.
33. Pui, C.H., et al., *Treating childhood acute lymphoblastic leukemia without cranial irradiation*. N Engl J Med, 2009. **360**(26): p. 2730-41.
34. Riley, P.A., *Free radicals in biology: oxidative stress and the effects of ionizing radiation*. Int J Radiat Biol, 1994. **65**(1): p. 27-33.
35. Cuscela, D., et al., *Protection from radiation-induced alopecia with topical application of nitroxides: fractionated studies*. Cancer J Sci Am, 1996. **2**(5): p. 273-8.
36. Ali, S.Y. and G. Singh, *Radiation-induced Alopecia*. Int J Trichology, 2010. **2**(2): p. 118-9.
37. Harrison, L.B., et al., *Radiotherapy-associated anemia: the scope of the problem*. Oncologist, 2000. **5 Suppl 2**: p. 1-7.
38. Muller, K. and V. Meineke, *Radiation-induced alterations in cytokine production by skin cells*. Exp Hematol, 2007. **35**(4 Suppl 1): p. 96-104.

39. Han, S.K., et al., *Effect of gamma radiation on cytokine expression and cytokine-receptor mediated STAT activation.* Int J Radiat Biol, 2006. **82**(9): p. 686-97.
40. Roscoe, J.A., et al., *Treatment of radiotherapy-induced fatigue through a nonpharmacological approach.* Integr Cancer Ther, 2005. **4**(1): p. 8-13.
41. Molteni, A., et al., *Control of radiation-induced pneumopathy and lung fibrosis by angiotensin-converting enzyme inhibitors and an angiotensin II type 1 receptor blocker.* Int J Radiat Biol, 2000. **76**(4): p. 523-32.
42. Shaw, J., et al., *The small-molecule TNF-alpha modulator, UTL-5g, reduces side effects induced by cisplatin and enhances the therapeutic effect of cisplatin in vivo.* J Exp Ther Oncol, 2011. **9**(2): p. 129-37.
43. Qian, J., et al., *TNF-a modulator as radioprotective agent for liver.* Proc Amer Assoc Cancer Res, 2005. **46**(Tumore Biology 13: Radiation Bilogy 2, Abstract #2000).
44. Raghunand, N., et al., *Plasmalemmal pH-gradients in drug-sensitive and drug-resistant MCF-7 human breast carcinoma xenografts measured by 31P magnetic resonance spectroscopy.* Biochem Pharmacol, 1999. **57**(3): p. 309-12.
45. Vaupel, P., *Hypoxia in neoplastic tissue.* Microvasc Res, 1977. **13**(3): p. 399-408.
46. Tannock, I.F. and D. Rotin, *Acid pH in tumors and its potential for therapeutic exploitation.* Cancer Res, 1989. **49**(16): p. 4373-84.
47. Praetorius, N.P. and T.K. Mandal, *Alternate delivery route for amifostine as a radio-/chemo-protecting agent.* J Pharm Pharmacol, 2008. **60**(7): p. 809-15.
48. Hausheer, F.H., et al., *Modulation of platinum-induced toxicities and therapeutic index: mechanistic insights and first- and second-generation protecting agents.* Semin Oncol, 1998. **25**(5): p. 584-99.
49. Gillio, A.P. and J.L. Gabrilove, *Cytokine treatment of inherited bone marrow failure syndromes.* Blood, 1993. **81**(7): p. 1669-74.
50. Marsh, J.C., A. Ganser, and M. Stadler, *Hematopoietic growth factors in the treatment of acquired bone marrow failure states.* Semin Hematol, 2007. **44**(3): p. 138-47.
51. Eschbach, J.W. and J.W. Adamson, *Anemia of end-stage renal disease (ESRD).* Kidney Int, 1985. **28**(1): p. 1-5.
52. Eschbach, J.W., et al., *Correction of the anemia of end-stage renal disease with recombinant human erythropoietin. Results of a combined phase I and II clinical trial.* N Engl J Med, 1987. **316**(2): p. 73-8.
53. Sola-Visner, M.C., et al., *Megakaryocyte size and concentration in the bone marrow of thrombocytopenic and nonthrombocytopenic neonates.* Pediatr Res, 2007. **61**(4): p. 479-84.
54. Kuter, D.J., et al., *Efficacy of romiplostim in patients with chronic immune thrombocytopenic purpura: a double-blind randomised controlled trial.* Lancet, 2008. **371**(9610): p. 395-403.
55. Cooper, N., I. Terrinoni, and A. Newland, *The efficacy and safety of romiplostim in adult patients with chronic immune thrombocytopenia.* Ther Adv Hematol, 2012. **3**(5): p. 291-8.
56. Broudy, V.C. and N.L. Lin, *AMG531 stimulates megakaryopoiesis in vitro by binding to Mpl.* Cytokine, 2004. **25**(2): p. 52-60.
57. Cashen, A.F., B. Nervi, and J. DiPersio, *AMD3100: CXCR4 antagonist and rapid stem cell-mobilizing agent.* Future Oncol, 2007. **3**(1): p. 19-27.

58. Lyman, G.H., N.M. Kuderer, and B. Djulbegovic, *Prophylactic granulocyte colony-stimulating factor in patients receiving dose-intensive cancer chemotherapy: a meta-analysis.* Am J Med, 2002. **112**(5): p. 406-11.
59. Lyman, G.H., D.C. Dale, and J. Crawford, *Incidence and predictors of low dose-intensity in adjuvant breast cancer chemotherapy: a nationwide study of community practices.* J Clin Oncol, 2003. **21**(24): p. 4524-31.
60. !!! INVALID CITATION !!!
61. T, E., et al., *DNA repair and mutagenesis.* 2006, Washington , D.C.: ASM Press.
62. Alberts, B., in *Molecular Biology of the Cells.* 2002, Garland Science: New York.
63. Nishino, T. and K. Morikawa, *Structure and function of nucleases in DNA repair: shape, grip and blade of the DNA scissors.* Oncogene, 2002. **21**(58): p. 9022-32.
64. FDA. *Ethyol (amifostine) For Injection, Package Insert.* Safety Labeling Changes Approved By FDA Center for Drug Evaluation and Research (CDER) -- November 2008 2008; Available from: http://www.fda.gov/Safety/MedWatch/SafetyInformation/Safety-RelatedDrugLabelingChanges/ucm121181.htm.
65. *Amifostine. USP DI, Drug information for the health care professional.* 2002: Englewood, Colorado: Micromedex, Inc.
66. *Alza Pharmaceuticals - US, Ethyol package insert,* 1996.
67. Korst, A.E., et al., *Pharmacokinetics of amifostine and its metabolites in the plasma and ascites of a cancer patient.* Cancer Chemother Pharmacol, 1996. **39**(1-2): p. 162-6.
68. Cavalli, F., et al., *A case of liver toxicity following cis-dichlorodiammineplatinum(II) treatment.* Cancer Treat Rep, 1978. **62**(12): p. 2125-6.
69. Cersosimo, R.J., *Hepatotoxicity associated with cisplatin chemotherapy.* Ann Pharmacother, 1993. **27**(4): p. 438-41.
70. Pollera, C.F., et al., *Cisplatin-induced hepatic toxicity.* J Clin Oncol, 1987. **5**(2): p. 318-9.
71. Culy, C.R. and C.M. Spencer, *Amifostine: an update on its clinical status as a cytoprotectant in patients with cancer receiving chemotherapy or radiotherapy and its potential therapeutic application in myelodysplastic syndrome.* Drugs, 2001. **61**(5): p. 641-84.
72. Mabro, M., S. Faivre, and E. Raymond, *A risk-benefit assessment of amifostine in cytoprotection.* Drug Saf, 1999. **21**(5): p. 367-87.
73. Shaw, J., et al., *The small-molecule TNF-alpha inhibitor, UTL-5g, delays deaths and increases survival rates for mice treated with high doses of cisplatin.* Cancer Chemother Pharmacol, 2013. **72**(3): p. 703-7.
74. Lunt, S.J., et al., *Tirapazamine administered as a neoadjuvant to radiotherapy reduces metastatic dissemination.* Clin Cancer Res, 2005. **11**(11): p. 4212-6.
75. Denny, W.A., *Prospects for hypoxia-activated anticancer drugs.* Curr Med Chem Anticancer Agents, 2004. **4**(5): p. 395-9.
76. Gandara, D.R., et al., *Tirapazamine: prototype for a novel class of therapeutic agents targeting tumor hypoxia.* Semin Oncol, 2002. **29**(1 Suppl 4): p. 102-9.
77. Itani, W., et al., *Radiosensitization by 2-benzoyl-3-phenyl-6,7-dichloroquinoxaline 1,4-dioxide under oxia and hypoxia in human colon cancer cells.* Radiat Oncol, 2007. **2**: p. 1.

78. Haykal, J., et al., *The radiosensitizer 2-benzoyl-3-phenyl-6,7-dichloroquinoxaline 1,4-dioxide induces DNA damage in EMT-6 mammary carcinoma cells.* Radiat Oncol, 2009. **4**: p. 25.
79. *A study of the effect of misonidazole in conjunction with radiotherapy for the treatment of grades 3 and 4 astrocytomas. A report from the MRC Working Party on misonidazole in gliomas.* Br J Radiol, 1983. **56**(669): p. 673-82.
80. Teicher, B.A., et al., *Cytotoxicity, radiosensitization, antitumor activity, and interaction with hyperthermia of a Co(III) mustard complex.* Cancer Res, 1990. **50**(21): p. 6971-5.
81. Noble, S. and J.A. Balfour, *Meloxicam.* Drugs, 1996. **51**(3): p. 424-30; discussion 431-32.
82. Bijnsdorp, I.V., et al., *Radiosensitizing potential of the selective cyclooygenase-2 (COX-2) inhibitor meloxicam on human glioma cells.* J Neurooncol, 2007. **85**(1): p. 25-31.
83. Demiral, A.N., et al., *Amifostine-induced toxic epidermal necrolysis during radiotherapy: a case report.* Jpn J Clin Oncol, 2002. **32**(11): p. 477-9.
84. Bogo, V., A.J. Jacobs, and J.F. Weiss, *Behavioral toxicity and efficacy of WR-2721 as a radioprotectant.* Radiat Res, 1985. **104**(2 Pt 1): p. 182-90.
85. Glover, D., et al., *WR-2721 protects against the hematologic toxicity of cyclophosphamide: a controlled phase II trial.* J Clin Oncol, 1986. **4**(4): p. 584-8.
86. Sagowski, C., et al., *Studies on the radioprotective potency of amifostine on salivary glands of rats during fractionated irradiation: acute and late effects.* Eur Arch Otorhinolaryngol, 2003. **260**(1): p. 42-7.
87. Murray, D., et al., *Protection by WR-3689 against gamma-ray-induced intestinal damage: comparative effect on clonogenic cell survival, mouse survival, and DNA damage.* Radiat Res, 1989. **120**(2): p. 339-51.
88. Barnes, J., et al., *N-Heterocyclic compounds as radioprotectors II. Derivatives of pyridine and pyrimidine containing thiol precursors.* Eu J of Med Chem, 1988. **23**(3): p. 211-216.
89. Vasil'eva, T. and N. Lisina, *Synthesis and radioprotective activity of functionally-substituted methyl(alkyl)sulfoxides.* Pharm Chem J, 1997. **31**: p. 593-596.
90. Kang, A.D., et al., *ON01210.Na (Ex-RAD(R)) mitigates radiation damage through activation of the AKT pathway.* PLoS One, 2013. **8**(3): p. e58355.
91. Citrin, D., et al., *Radioprotectors and mitigators of radiation-induced normal tissue injury.* Oncologist, 2010. **15**(4): p. 360-71.
92. Ahmad, S., et al., *Transforming growth factor-beta negatively modulates T-cell responses in sepsis.* FEBS Lett, 1997. **402**(2-3): p. 213-8.
93. Yeh, F.L., H.D. Shen, and R.H. Fang, *Deficient transforming growth factor beta and interleukin-10 responses contribute to the septic death of burned patients.* Burns, 2002. **28**(7): p. 631-7.
94. Rigato, O., et al., *Tumor necrosis factor alpha (TNF-alpha) and sepsis: evidence for a role in host defense.* Infection, 1996. **24**(4): p. 314-8.
95. Marie, C., J.M. Cavaillon, and M.R. Losser, *Elevated levels of circulating transforming growth factor-beta 1 in patients with the sepsis syndrome.* Ann Intern Med, 1996. **125**(6): p. 520-1.
96. Weehuizen, T.A., et al., *Expression and function of transforming growth factor beta in melioidosis.* Infect Immun, 2012. **80**(5): p. 1853-7.

97. Maisin, J.R., A. Kondi-Tamba, and G. Mattelin, *Polysaccharides induce radioprotection of murine hemopoietic stem cells and increase the LD50/30 days.* Radiat Res, 1986. **105**(2): p. 276-81.
98. Patchen, M.L., et al., *Radioprotection by polysaccharides alone and in combination with aminothiols.* Adv Space Res, 1992. **12**(2-3): p. 233-48.
99. Maxhimer, J., et al., *Radioprotection in Normal Tissue and Delayed Tumor Growth by Blockade of CD47 Signaling* Science Translational Medicine 2009. **1**(3): p. 3.
100. Huang, X.W., et al., *Antisense oligonucleotide inhibition of tumor necrosis factor receptor 1 protects the liver from radiation-induced apoptosis.* Clin Cancer Res, 2006. **12**(9): p. 2849-55.
101. Uematsu, S., et al., *Detection of pathogenic intestinal bacteria by Toll-like receptor 5 on intestinal CD11c+ lamina propria cells.* Nat Immunol, 2006. **7**(8): p. 868-74.
102. Burdelya, L.G., et al., *An agonist of toll-like receptor 5 has radioprotective activity in mouse and primate models.* Science, 2008. **320**(5873): p. 226-30.
103. Topkan, E., et al., *Comparison of the protective effects of melatonin and amifostine on radiation-induced epiphyseal injury.* Int J Radiat Biol, 2008. **84**(10): p. 796-802.
104. Hussein, M.R., et al., *Melatonin and roentgen irradiation-induced acute radiation enteritis in Albino rats: an animal model.* Cell Biol Int, 2008. **32**(11): p. 1353-61.
105. Manda, K., M. Ueno, and K. Anzai, *Cranial irradiation-induced inhibition of neurogenesis in hippocampal dentate gyrus of adult mice: attenuation by melatonin pretreatment.* J Pineal Res, 2009. **46**(1): p. 71-8.
106. Blask, D., et al., *Melatonin radiosensitizes tumors and radioprotects normal tissues: Time-of day dependency.* AmAssoc Cancer Res 2000. **Abstract 338**.
107. Hsu, W.-L., et al., *Is melanin a radioprotector or radiosensitizer? It's implication for radiotherapy.* Therapeut Radiol Oncol 2003. **10**(4): p. 237-246.
108. Zhang, Q., et al., *Novel quinic acid derivative KZ-41 prevents retinal endothelial cell apoptosis without inhibiting retinoblastoma cell death through p38 signaling.* Invest Ophthalmol Vis Sci, 2013. **54**(9): p. 5937-43.
109. Rube, C.E., et al., *Modulation of radiation-induced tumour necrosis factor alpha (TNF-alpha) expression in the lung tissue by pentoxifylline.* Radiother Oncol, 2002. **64**(2): p. 177-87.
110. Redlich, C.A., et al., *IL-11 enhances survival and decreases TNF production after radiation-induced thoracic injury.* J Immunol, 1996. **157**(4): p. 1705-10.
111. Chen, M.F., et al., *Caffeic acid phenethyl ester decreases acute pneumonitis after irradiation in vitro and in vivo.* BMC Cancer, 2005. **5**: p. 158.
112. Shaw, J., et al., *Anti-inflammatory and Anti-arthritic Effects of a Novel Leflunomide Analogue, UTL-5b (GBL-5b).* Am J Biomed Sci, 2011. **3**(1): p. 31-39.
113. Shaw, J., et al., *The small-molecule TNF-a modulator, UTL-5g, reduces side effects induced by cisplatin and enhances the therapeutic effects of cisplatin in vivo.* J Exp Ther Oncol (in press), 2011.
114. McCumber, L.M., *The potential influence of cell protectors for dose escalation in cancer therapy: an analysis of amifostine.* Med Dosim, 2004. **29**(2): p. 139-43.

115. Brown, S., et al., *UTL-5g lowers levels of TGF-β and TNF-a elevated by lung irradiation and does not affect tumor-response to irradiation.* Am J Biomed Sci, 2014. **6**(3): p. 157-165.
116. Zeng, K., et al., *Preclinical pharmacokinetics of the radiomitigator KZ-41 in rats.* Xenobiotica, 2011. **41**(11): p. 1006-12.
117. Toutounchian, J.J., et al., *Modulation of radiation injury response in retinal endothelial cells by quinic acid derivative KZ-41 involves p38 MAPK.* PLoS One, 2014. **9**(6): p. e100210.
118. Thompson, K.E., et al., *Quinic acid derivative KZ-41 exhibits radiomitigating activity in preclinical models of radiation injury.* Drug Dev Res, 2014. **75**(1): p. 29-36.
119. Brines, M. and A. Cerami, *Erythropoietin-mediated tissue protection: reducing collateral damage from the primary injury response.* J Intern Med, 2008. **264**(5): p. 405-32.
120. Yavas, G., et al., *The effect of Halofuginone in the amelioration of radiation induced-lung fibrosis.* Med Hypotheses, 2013. **80**(4): p. 357-9.
121. Xavier, S., et al., *Amelioration of radiation-induced fibrosis: inhibition of transforming growth factor-beta signaling by halofuginone.* J Biol Chem, 2004. **279**(15): p. 15167-76.
122. Tzurel, A., et al., *Halofuginone does not reduce fibrosis in bleomycin-induced lung injury.* Life Sci, 2002. **71**(14): p. 1599-606.
123. Konopacka, M., M. Widel, and J. Rzeszowska-Wolny, *Modifying effect of vitamins C, E and beta-carotene against gamma-ray-induced DNA damage in mouse cells.* Mutat Res, 1998. **417**(2-3): p. 85-94.
124. Wadleigh, R.G., et al., *Vitamin E in the treatment of chemotherapy-induced mucositis.* Am J Med, 1992. **92**(5): p. 481-4.
125. Kottschade, L., et al., *The use of vitamin E for prevention of chemotherapy-induced peripheral neuropathy: A phase III double-blind, placebo controlled study – N05C3.* J Clin Oncol, 2009. **27**: p. 15s.
126. Kumar, B., et al., *D-alpha-tocopheryl succinate (vitamin E) enhances radiation-induced chromosomal damage levels in human cancer cells, but reduces it in normal cells.* J Am Coll Nutr, 2002. **21**(4): p. 339-43.
127. Prasad, K.N., W.C. Cole, and K.C. Prasad, *Risk factors for Alzheimer's disease: role of multiple antioxidants, non-steroidal anti-inflammatory and cholinergic agents alone or in combination in prevention and treatment.* J Am Coll Nutr, 2002. **21**(6): p. 506-22.
128. Ahmadi, A., et al., *Chemoprotective effects of hesperidin against genotoxicity induced by cyclophosphamide in mice bone marrow cells.* Arch Pharm Res, 2008. **31**(6): p. 794-7.
129. Michelson, A., *Superoxide and Superoxide Dismutase.* Toxicity of superoxide radical anions, ed. A. Michelson, I. McCord, and I. Fridovich. 1977: Academic Press.
130. Klug, D., J. Rabani, and I. Fridovich, *A direct demonstration of the catalytic action of superoxide dismutase through the use of pulse radiolysis.* J Biol Chem, 1972. **247**(15): p. 4839-42.
131. Petkau, A. and W.S. Chelack, *Radioprotection by superoxide dismutase of macrophage progenitor cells from mouse bone marrow.* Biochem Biophys Res Commun, 1984. **119**(3): p. 1089-95.
132. Jones, J.B., et al., *Radioprotective effect of free radical scavenging enzymes.* J Otolaryngol, 1990. **19**(5): p. 299-306.

133. Overgaard, J., et al., *Lack of radiation protective effect of orgotein in normal and malignant mammalian cells.* Acta Radiol Oncol Radiat Phys Biol, 1979. **18**(4): p. 305-12.
134. Marenzi, G., et al., *N-acetylcysteine and contrast-induced nephropathy in primary angioplasty.* N Engl J Med, 2006. **354**(26): p. 2773-82.
135. Devipriya, N., et al., *Quercetin ameliorates gamma radiation-induced DNA damage and biochemical changes in human peripheral blood lymphocytes.* Mutat Res, 2008. **654**(1): p. 1-7.
136. Liu, W.C., et al., *Protection against radiation-induced bone marrow and intestinal injuries by Cordyceps sinensis, a Chinese herbal medicine.* Radiat Res, 2006. **166**(6): p. 900-7.
137. Devi, P.U., K.S. Bisht, and M. Vinitha, *A comparative study of radioprotection by Ocimum flavonoids and synthetic aminothiol protectors in the mouse.* Br J Radiol, 1998. **71**(847): p. 782-4.
138. Rahman, S. and S. Sultana, *Glycyrrhizin exhibits potential chemopreventive activity on 12-O-tetradecanoyl phorbol-13-acetate-induced cutaneous oxidative stress and tumor promotion in Swiss albino mice.* J Enzyme Inhib Med Chem, 2007. **22**(3): p. 363-9.
139. Greenlee, H., et al., *Clinical applications of Silybum marianum in oncology.* Integr Cancer Ther, 2007. **6**(2): p. 158-65.
140. Jayaraj, R., et al., *Hepatoprotective efficacy of certain flavonoids against microcystin induced toxicity in mice.* Environ Toxicol, 2007. **22**(5): p. 472-9.
141. Hogan, F.S., et al., *Flavonoid, silibinin, inhibits proliferation and promotes cell-cycle arrest of human colon cancer.* J Surg Res, 2007. **143**(1): p. 58-65.
142. Tamayo, C. and S. Diamond, *Review of clinical trials evaluating safety and efficacy of milk thistle (Silybum marianum [L.] Gaertn.).* Integr Cancer Ther, 2007. **6**(2): p. 146-57.
143. Szilard, S., D. Szentgyorgyi, and I. Demeter, *Protective effect of Legalon in workers exposed to organic solvents.* Acta Med Hung, 1988. **45**(2): p. 249-56.
144. Yuen, J.W. and M.D. Gohel, *Anticancer effects of Ganoderma lucidum: a review of scientific evidence.* Nutr Cancer, 2005. **53**(1): p. 11-7.
145. Xu, Z., et al., *Ganoderma lucidum polysaccharides: immunomodulation and potential anti-tumor activities.* Am J Chin Med, 2011. **39**(1): p. 15-27.
146. Sanodiya, B.S., et al., *Ganoderma lucidum: a potent pharmacological macrofungus.* Curr Pharm Biotechnol, 2009. **10**(8): p. 717-42.
147. Samarth, R.M. and M. Samarth, *Protection against radiation-induced testicular damage in Swiss albino mice by Mentha piperita (Linn.).* Basic Clin Pharmacol Toxicol, 2009. **104**(4): p. 329-34.
148. Baliga, M.S. and S. Rao, *Radioprotective potential of mint: a brief review.* J Cancer Res Ther, 2010. **6**(3): p. 255-62.

Index

Made in the USA
Monee, IL
16 September 2026

63312026R00079